THE

EVERYTHING® GUIDE TO CALCULUS I

Dear Reader,

Ever since my initial introduction to calculus as a high school senior in 1974, I have been fascinated by the subject. Looking back over my thirty years as a high school mathematics teacher provides an even more interesting perspective on how the teaching and learning of calculus have evolved even in such a relatively short period of time. I first learned calculus without the benefit of technological supports. The course consisted of a great deal of memorization of definitions, theorems, and rules applied to very algebraically complex problems. It was not uncommon for me to solve a problem and not really understand what I had just accomplished.

In the past two decades, graphing technologies, computer software, and Internet applets have changed the way calculus is taught and understood. The factual information is still the same, but students can now view local linearity, watch an applet turn a secant line into a tangent line, and see the number of inscribed rectangles increase to produce increasingly better approximations of areas under graphs. The beautiful geometric principles in the course are more salient, and realistic applications are more available. I hope your experience with this book inspires in you a similar passion for this wonderful subject.

Sincerely,

Greg Hill

Welcome to the EVERYTHING® Series!

These handy, accessible books give you all you need to tackle a difficult project, gain a new hobby, comprehend a fascinating topic, prepare for an exam, or even brush up on something you learned back in school but have since forgotten.

You can choose to read an Everything® book from cover to cover or just pick out the information you want from our four useful boxes: e-rules, e-questions, e-alerts, and e-ssentials. We give you everything you need to know on the subject, but throw in a lot of fun stuff along the way, too.

We now have more than 400 Everything® books in print, spanning such wide-ranging categories as weddings, pregnancy, cooking, music instruction, foreign language, crafts, pets, New Age, and so much more. When you're done reading them all, you can finally say you know Everything®!

RULE
Important rules to remember

QUESTION
Answers to common questions

ALERT
Urgent warnings

ESSENTIAL
Quick handy tips

PUBLISHER Karen Cooper
DIRECTOR OF ACQUISITIONS AND INNOVATION Paula Munier
MANAGING EDITOR, EVERYTHING® SERIES Lisa Laing
COPY CHIEF Casey Ebert
ASSISTANT PRODUCTION EDITOR Jacob Erickson
ACQUISITIONS EDITOR Lisa Laing
ASSOCIATE DEVELOPMENT EDITOR Hillary Thompson
EDITORIAL ASSISTANT Ross Weisman
EVERYTHING® SERIES COVER DESIGNER Erin Alexander
LAYOUT DESIGNER Erin Dawson
ILLUSTRATOR Evan Hill

THE EVERYTHING® GUIDE TO Calculus I

A step-by-step guide to the basics of calculus—*in plain English!*

Greg Hill
National Council of Teachers of Mathematics

This book is dedicated to my good friends and colleagues John Brunsting and John Diehl, two of my most significant mentors in mathematics—and particularly in the field of calculus.

Published by Adams Media, a division of F+W Media, Inc.
57 Littlefield Street, Avon, MA 02322 U.S.A.
www.adamsmedia.com

ISBN 10: 1-4405-0629-9
ISBN 13: 978-1-4405-0629-1
eISBN 10: 1-4405-0630-2
eISBN 13: 978-1-4405-0630-7

Printed in the United States of America.

10 9 8 7 6 5 4 3 2 1

Library of Congress Cataloging-in-Publication Data
is available from the publisher.

Contents

Acknowledgments

Thank you to Cathy, my wonderful wife of twenty-five years, who always supports and encourages me in each professional endeavor. Thank you also to my son Evan for the many hours he spent producing all of the artwork for this book. I am also very appreciative of the steady support of my managing editor Lisa. When I was unsure whether a book about calculus would fit the Everything® series format, Lisa instilled the confidence to move forward with the challenge. She also kindly provided the impetus to see it through to completion. Finally, thank you to my friends and colleagues in the math department at Hinsdale Central High School, where I have taught for thirty years. They have been excited about and interested in this project each step of the way, and their enthusiasm has been extremely uplifting during many long hours of writing.

The Top 10 Ways to Be Successful in Calculus

1. Develop a solid foundation in algebra, geometry, and trigonometry.
2. Seek a balance between analytic skills and conceptual understanding. Understanding the big ideas makes mastering the mechanics much easier.
3. Memorize the facts that are the foundation of the course, including derivative and integral rules and major theorems.
4. Think about most concepts on a microscopic level. Calculus is the study of change on infinitely small intervals and how those changes accumulate.
5. Remember the chain rule. It shows up in almost every derivative and integral.
6. Use visualization tools to make connections between ideas that are important in the subject. Graphing technologies and Internet applets are excellent sources of such tools.
7. Be patient. Master each concept before moving ahead, because the subject builds upon itself incrementally.
8. Read and study from multiple sources. A single explanation of an idea may not make it clear, and there are numerous paths to understand any concept.
9. Find a friend or mentor to discuss the ideas that arise in the course.
10. Look for the beauty and wonder in this subject! Change is everywhere in the world, and calculus helps us analyze and understand it.

Introduction

FOR DECADES, CALCULUS has struck fear into the hearts of countless high school seniors and first-year college students. The mere thought of mastering the mysteries of the subject has caused far too many people to give up before they even get started. But for those who gain an understanding of it, calculus is a beautiful integration (no pun intended) of all the math topics that lead up to it. Calculus uses arithmetic, algebra, geometry, and trigonometry to develop new and fascinating ideas. The most important thing to do as you work through this book is to really believe that you can learn calculus. Its reputation as an unconquerable mountain is totally undeserved.

Centuries ago, the mathematicians Isaac Newton, Gottfried Leibniz, Leonhard Euler, and others worked to develop the ideas of calculus in an attempt to study and understand the world around them. At its most basic, calculus is the mathematics of change. These mathematicians realized that by thinking in terms of infinitely small increments, they could better understand ideas of limits, rates of change, and even areas and volumes of irregularly shaped regions and objects.

Today, calculus is a vital element in the foundation of many practical fields, such as engineering, biological sciences, medical studies, economics, and even the automobile and film industries. People in these professions are not necessarily sitting down at their desks and working calculus problems. It is more likely that the tools they use, particularly computer programs, have calculus processes at their core. Those tools are used to achieve an end or produce a result, and without the help of calculus, the tasks would be significantly more difficult—and sometimes even impossible.

This book is intended to remove the mystery of learning calculus. You will see how calculus utilizes many of the foundational ideas introduced in earlier courses of study to develop new ideas. You will also discover that, taken in small steps and developed gradually, the big ideas of calculus are very accessible. It is not the evil monster many people make it out to be. The book will give the subject relevance by helping you understand how the calculus concept of measuring and exploring summations of infinitely small change shed light on what is happening at any instant, what has happened in the past, and what may happen in the near future. Currently, global warming is a major concern for environmentalists. Scientists are constantly studying changes in the Earth's temperature. Those changes are examined over short intervals to establish current rates of change, and they are studied over longer intervals to get a picture of the long-term impact of the phenomenon. Although it is cloaked here in environmental science, this is the essence of calculus: the study of change.

Ironically, calculus itself continues to change in certain respects. The capabilities of computers and calculators have opened new pathways to understanding the ideas of calculus. Graphing calculators and computer algebra systems reduce some of the challenges that students faced in past decades. Graphing calculators can rapidly produce graphs, solve equations, and even numerically evaluate derivatives and integrals. Computer algebra systems can significantly reduce the manipulations necessary for solving complicated equations, taking numerous derivatives, or finding antiderivatives. Geometry software programs make possible the dynamic visualization of many calculus concepts. Used properly as tools of study, these newer technologies can reduce the mechanical obstacles and sharpen the focus on the big ideas of the course.

CHAPTER 1

Prerequisite Skills

With any new endeavor, you usually need to have certain basic skills to move ahead successfully. Could you imagine becoming a triathlete if you didn't know how to swim or ride a bike? The same is true for calculus. All the math courses you've taken create a solid foundation for calculus. Don't believe all the hype about how hard calculus is. If you've taken algebra, geometry, and trigonometry, you've got everything you need to get started.

Important Algebra Skills

Some people say that by the end of middle school, students have learned all of the math they need to know to get by in life. That may be true for many people, but for those who want to build a career using any kind of math skill, a mastery of basic algebra is indispensable in all successive math courses.

Algebra introduces abstract thinking into the world of numbers and equations via the use of variables to represent unknown quantities. Algebra students learn how to use the Cartesian coordinate system to view graphs and data. Many students first learn to use a graphing calculator when they study algebra, and this skill is a huge part of calculus. Many other things you learned in algebra, such as pattern finding, variable expressions and functions, powers of variables, and properties of exponents, are used regularly in calculus. You'll recognize polynomials and factoring here, as well as domain and range. And you've already learned perhaps the most important tool for working in calculus: solving equations and inequalities.

This chapter is a quick review of these concepts from algebra class. You won't need to go over everything you learned in Algebra I and II, but you'll concentrate on the algebra building blocks you'll need. Remember, as you're working through the problems in this book, you can refer to the key algebra formulas in Appendix A.

Solving Equations

When you solve an equation, you find the value or values of the variables that make that equation true. You first simplify the expressions on either side of the equation. As you do this, keep the equation balanced by always doing the same mathematical step to both sides of the equation. Look at the example that follows. It's a pretty simple first-degree equation. Do you remember those?

EXAMPLE 1-1

Solve $3(x-1)-7=\frac{x}{2}$.

Simplify the left-hand side by distributing the 3. $3x-3-7=\frac{x}{2}$

Combine the constants. $3x-10=\frac{x}{2}$

Multiply both sides of the equation by 2. $6x-20=x$

Subtract $6x$ from both sides of the equation. $-20=-5x$

Divide both sides of the equation by –5. $4=x$

Solving Inequalities

In calculus, you'll often need to determine where a function is zero, positive, or negative. This may happen at specific numbers or over a whole set of numbers, so you must pay attention to the domain of the problem. Changing the previous example into an inequality problem should remind you of several important ideas about solving inequalities and about solving over a certain domain.

A small detail in solving inequalities makes a big difference between getting the right answer and getting the wrong answer. If you multiply or divide both sides of an inequality by a negative number, you must turn the inequality arrow around.

Suppose the problem were to solve $3(x-1)-7<\frac{x}{2}$.

The initial steps would have been similar up to the point where $-20<-5x$. When you now divide both sides of the equation by –5, the inequality symbol turns around to give $4>x$. If no domain for the solution or solutions is stated or can be determined from the context of the problem, you should assume that the domain is all real numbers. In this case, every number less than 4 would solve the original problem. But if the problem had been stated as "Solve $3(x-1)-7<\frac{x}{2}$ over the whole numbers," the only acceptable solutions would have been the whole numbers less than 4, which are 0, 1, 2, and 3.

Graphing on the Cartesian Coordinate System

Don't let the fancy term intimidate you. The Cartesian coordinate system, named for the French mathematician René Descartes, is just the xy-plane in which all graphing occurs. The graphs of two basic types of functions, linear and quadratic, are often introduced in algebra, and they resurface frequently in further math. Calculus, a very visual subject, requires working with quite a few graphs of lines, parabolas, and other functions. Most often, you will have the equations and will need to produce a quick sketch, but at times, you may need to write the equation for a line given certain information.

The slope-intercept form of a linear equation has the form $y = mx + b$, where m is the slope and b is the y-intercept. When the power on the x is 1, the graph will always produce a line, which is why the equation is called *linear.*

One way to graph any function is to pick a few x values, plug them into the equation to calculate the corresponding y values, and then plot the points and connect them. Because many graphs used in calculus are just tools employed to analyze a problem or work toward a solution, a quick sketch is often sufficient. Generating individual points can be too time-consuming. Let's look at a better approach.

The slope of a line (the m in $y = mx + b$) is the ratio of how much the y values change to how much the x values change. If you know the slope, you have an easy way to move from point to point on the graph. The b value is the y-intercept because when $x = 0$, then $y = b$. This is where the graph will cross the y-axis. Therefore, a much more efficient way to produce the graph of a line is to plot a point at the y-intercept and then use the slope to find other points on the line before connecting them. All linear equations, with the exception of vertical lines, can be rearranged into slope-intercept form, so this quick sketch method can almost always be used.

EXAMPLE 1-2

Graph $y = \frac{3}{2}x - 2$.

Plot a point at $(0,-2)$ because that is the y-intercept. From that point, move up 3 units and right 2 units to find another point on the line. You can also move down 3 units and left 2 units because $\frac{3}{2}=\frac{-3}{-2}$. Connect the points with a straight line.

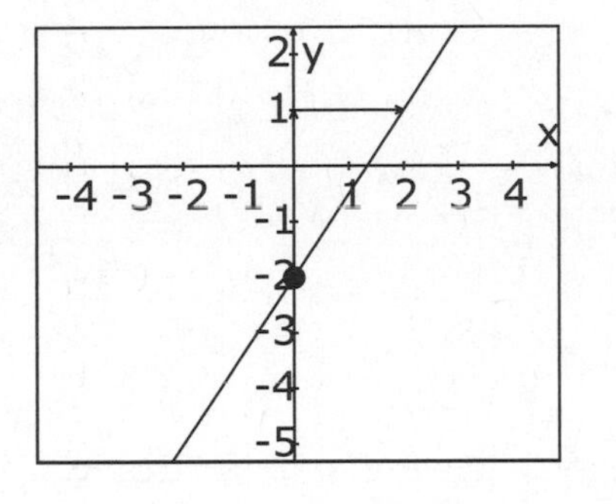

Figure 1-1

Graphing quadratic equations is somewhat more involved, but remember that in calculus, a rough sketch, with knowledge of a few key characteristics of the function, is frequently all the information you need. It's often enough to know the location of the vertex, the y-intercept, and the x-intercepts of the graph, if they exist. Where the graph lies in relation to the x-axis will provide important information—specifically, where the function values are positive and where they are negative.

The standard form of a quadratic equation is $y=a\cdot x^2+b\cdot x+c$, where a, b, and c are all real numbers. Quadratic equations produce U-shaped graphs called parabolas. The x-coordinate of the vertex is found by $x=\frac{-b}{2a}$. The y-coordinate is found by substituting the calculated x value into the equation. From $y=a\cdot x^2+b\cdot x+c$, the y-intercept is c because when $x=0,\ y=c$.

EXAMPLE 1-3

Produce a sketch of the quadratic function $y = 2x^2 - 4x - 3$.

By comparison to standard form $y = a \cdot x^2 + b \cdot x + c$, $a = 2$, $b = -4$, and $c = -3$.

Using $x = \frac{-b}{2a}$ reveals that the x-coordinate of the vertex is $x = \frac{-(-4)}{2 \cdot 2}$, or $x = 1$.

The y-coordinate is $y = 2 \cdot 1^2 - 4 \cdot 1 - 3$, or $y = -5$.

Because a is a positive number, the parabola is an upward-oriented U-shape.

The y-intercept is c, which is -3.

Even without the x-intercepts, you should now be able to produce a quick sketch of the graph.

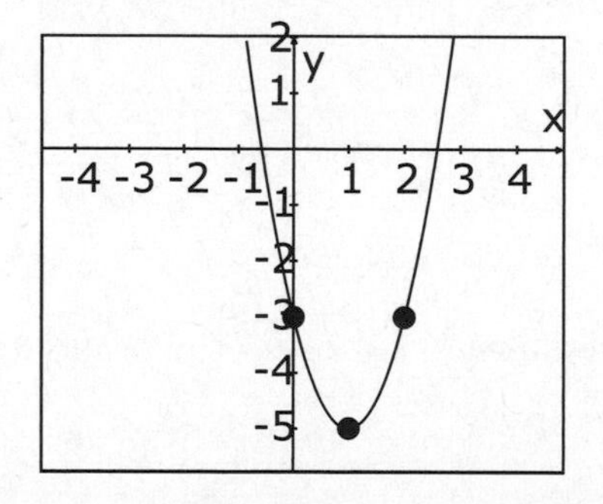

Figure 1-2

Note the point plotted at $(2, -3)$. Remember that all parabolas are symmetric, so you can easily find an additional point by using symmetry to the y-intercept.

Factoring Polynomial Expressions

The factoring you learned in algebra is used over and over again in successive math courses, and this is certainly true for calculus. The most common factoring you'll do in calculus is factoring the greatest common factor (GCF) out of an expression. You'll also use factoring to break a quadratic polynomial into the product of two simple binomials in order to solve an equation. Taking out the GCF means you essentially reverse the distributive property. Factoring a quadratic expression into a product of two binomials is a bit more complicated. To find the GCF of a polynomial, first find the GCF of all coefficients. For the variables in the expression, choose the lowest power of each variable that appears in all terms. To factor out the greatest common factor, divide each term of the original polynomial by the GCF.

EXAMPLE 1-4

Factor the greatest common factor out of $12x^3y^2 - 18x^4y^3z + 30x^4y^5$.

By observation, you should be able to tell that the greatest common factor of 12, 18, and 30 is 6. This is the first factor in your GCF.

The lowest power of x that is in each term is x^3.

The lowest power of y in all terms is y^2.

Because z does not appear in all terms, it is not part of the greatest common factor.

The GCF is $6x^3y^2$, and you must divide it out of the original polynomial to find the remaining factor.

$$\frac{12x^3y^2 - 18x^4y^3z + 30x^4y^5}{6x^3y^2} = 2 - 3xyz + 5xy^3$$

The final factored form is
$12x^3y^2 - 18x^4y^3z + 30x^4y^5 = 6x^3y^2(2 - 3xyz + 5xy^3)$.

In calculus, factoring a quadratic expression is used almost exclusively to find the solutions to an equation and to determine where the quadratic is positive, zero, or negative. Remember that a quadratic is usually the result of multiplying together two binomials. For example, $(x-3)(2x+9)=2x^2+3x-27$. Factoring takes the expanded quadratic and breaks it back down into its two factors. Do you remember an acronym called FOIL from algebra class? It stands for **F**irst, **O**utside, **I**nside, **L**ast.

Figure 1-3

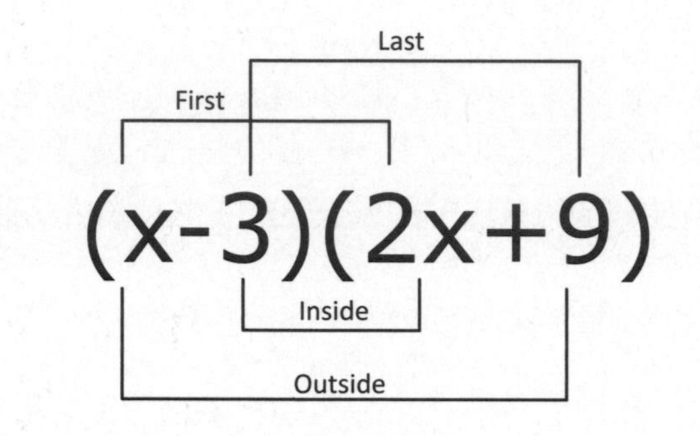

This is one way in which two binomials can be multiplied. The $2x^2$ above is the result of multiplying the first term in one of these binomials by the first term in the other: x times $2x$. The "outside" terms are x and 9, and the "inside" terms are −3 and $2x$. If you multiply each pair, you get $9x$ and $-6x$, which sum to $3x$. The last term in each binomial is −3 and 9. Their product is −27. When you factor a binomial, concentrate on getting the product of the first terms and the product of the last terms correct, and then check the middle term by summing the "outside" and "inside" products.

EXAMPLE 1-5

Factor x^2-5x+6.

Get the first terms in place to produce x^2. $(x \quad)(x \quad)$

Try the last terms to produce 6. $(x \quad - \quad 1)(x \quad - \quad 6)$

This doesn't work, because the middle term would be $-6x+(-1x)=-7x$.

Try other numbers that multiply to 6. $(x - 2)(x - 3)$

Note that the outside product plus the inside product is $-3x+(-2x)=-5x$.

Thus x^2-5x+6 factors into $(x - 2)(x - 3)$.

The Geometry of Calculus

There is a fairly common problem in most calculus courses that goes something like this: "A piece of steel 20 feet long and 6 feet wide is to be shaped into a long trough. It can be bent into an isosceles triangle, an isosceles trapezoid, or a semicircle. Which shape will produce the trough of greatest volume?" You don't need to solve this problem right now, but let's look at how geometry is involved in problems such as this.

In all possible cases, the trough will be in the shape of a prism. The volume of a prism is found by multiplying the area of its base times its length. The length is simply 20 feet, but to find the area of the base, you will need to know area formulas for all three possible shapes: triangle, semicircle, and trapezoid. You'll also need to know the angle at which you might bend the sides to form the triangle or the trapezoid. This could even introduce some trigonometry into the problem. There are several places where geometry concepts surface in calculus, and a quick survey of those concepts will prepare you to apply them when the need arises.

Area and Volume Formulas

The most common geometry formulas you will need to know involve area. Throughout calculus, shapes such as rectangles, triangles, trapezoids, and circles make frequent appearances. You'll also need to recall some volume formulas, especially the formulas for cylinders, washer-shaped objects, and spheres. You probably won't encounter more obscure geometry formulas very often, but you can look them up as needed. Common area and volume formulas are listed in Appendix A.

The most famous formula from geometry should always be at your fingertips for any calculus application. The Pythagorean Theorem expresses the relationship between the sides of a right triangle. It tells us that the sum of the squares of the two legs equals the square of the hypotenuse, the side across from the right angle: $(\text{Leg})^2 + (\text{Leg})^2 = (\text{Hypotenuse})^2$.

Tangents and Secants

In geometry, tangents and secants are associated primarily with circles. A *secant* is any line or segment that cuts across a circle, intersecting it at two points. A *tangent* line intersects a circle at only one point. In calculus, the context of these two terms is expanded and applied to graphs of functions. A line can be tangent to the graph of a function at a given point, but then intersect that function again at another point. Often, the context of the situation will be dealing with what is happening over a very local domain.

Special Right Triangles

Two special right triangles that you learned about in geometry should be reviewed. An isosceles right triangle has both legs congruent, which results in the angle measures being $45°, 45°$, and $90°$. What is more important is that the sides have a ratio of x to x to $x\sqrt{2}$. The other special right triangle has angles that measure $30°$, $60°$, and $90°$. Its sides are in a ratio of $2x$ to x to $x\sqrt{3}$. Figure 1-4 shows the relationships visually. It is very helpful to be familiar with these two triangles when you work with trigonometric values of common angle or radian measures.

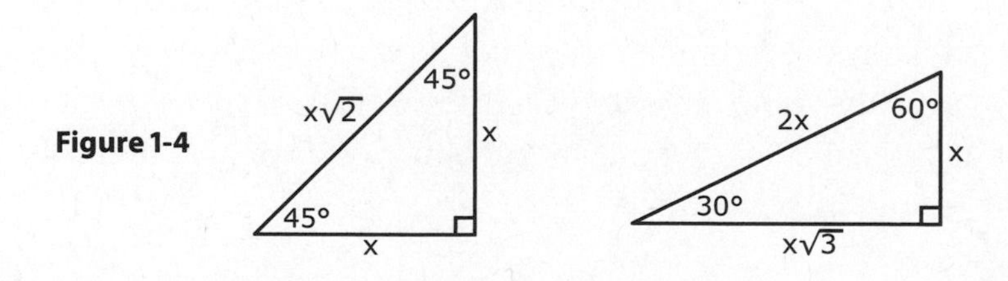

Figure 1-4

A Bit of Trigonometry

What is the exact area between the graph of $y = \sin(x)$ and the x-axis on the domain $\left[\frac{\pi}{6}, \frac{\pi}{3}\right]$? This is the kind of question that calculus can help you answer, but there are some basic trigonometry skills you will need while working on the task. Obviously, you will need to know what the graph of $y = \sin(x)$ looks like. You will also need to know some trigonometric values at the endpoints of the given interval. In general, most studies of trigonometry go into far more depth than is required for basic calculus. Of course, a calculus problem could take you to the outer limits of a trigonometry course, but for the most part, the trigonometric relationships that emerge throughout a calculus course are the simpler ones. What, then, should you remember?

Right Triangle Trigonometry

The basic definitions of the six trigonometric functions are expressed using a right triangle. The six functions are:

- Sine
- Cosine
- Tangent
- Cotangent
- Secant
- Cosecant

Each function has its own definition. For instance, the sine function for an angle is always the ratio of the length of the leg opposite that angle to the length of the hypotenuse. To describe all six would get rather wordy. Instead, study the triangle and the definitions in Figure 1-5 to review the basic trigonometric ratios.

Figure 1-5

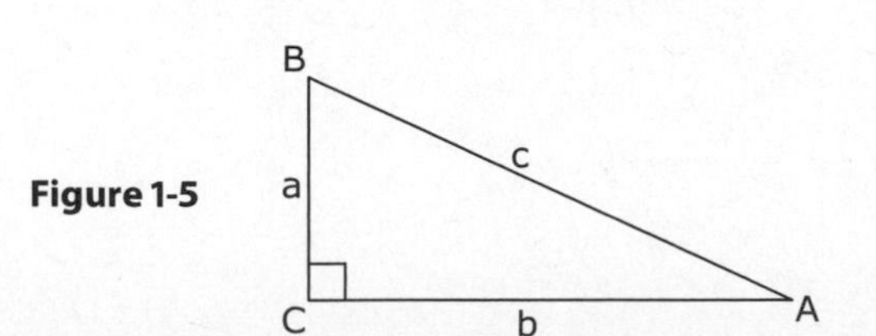

$$\sin(\angle A) = \frac{a}{c} \quad \cos(\angle A) = \frac{b}{c} \quad \tan(\angle A) = \frac{a}{b}$$

$$\csc(\angle A) = \frac{c}{a} \quad \sec(\angle A) = \frac{c}{b} \quad \cot(\angle A) = \frac{b}{a}$$

Note that sine and cosecant are reciprocals of each other. The same is true for cosine and secant, as well as for tangent and cotangent.

Common Unit Circle Values

Because the trigonometric values are defined as ratios, the size of the right triangle does not matter. A 30° angle in a right triangle of any size will always have a sine value of $\frac{1}{2}$. It is often easiest to work on right triangles with a hypotenuse of length 1 unit, so mathematicians established the unit circle.

A location on the unit circle is arrived at by rotating a certain number of degrees or radians in either direction from the point (1, 0). Counterclockwise rotation defines a positive angle, and clockwise rotation defines a negative angle. The *x*- and *y*-coordinates on the unit circle at the ending point, or terminal point, are respectively the cosine and sine of the rotation angle that lands you there.

In calculus you will frequently have to evaluate a trigonometric value of some common angle or radian measure. Figure 1-6 shows a unit circle with two rotation angles. Seeing the orientation of the special right triangles helps you find the coordinates.

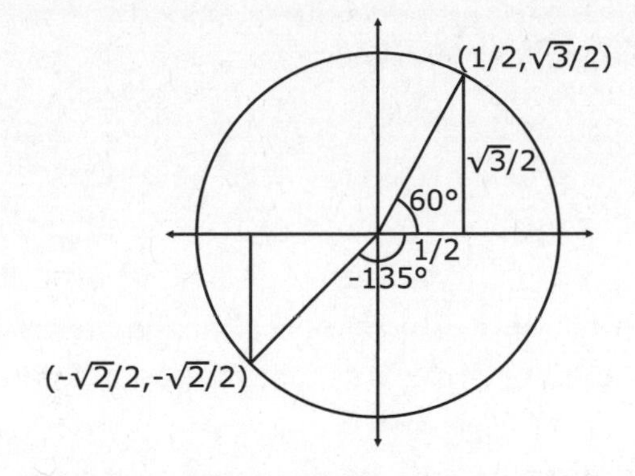

Figure 1-6

EXAMPLE 1-6

Use Figure 1-6 to answer the next three questions.

1. Find the exact value of $\sin(60°)$.

2. Find the exact value of $\cos(-135°)$.

3. Find the exact value of $\tan(225°)$.

SOLUTIONS

1. The y-coordinate at 60° is $\frac{\sqrt{3}}{2}$.

2. The x-coordinate at –135° is $-\frac{\sqrt{2}}{2}$.

3. A positive rotation of 225° lands you in the same location as –135°.

On the unit circle, the tangent is the y-coordinate divided by the x-coordinate: $\tan(225°) = \frac{-\sqrt{2}/2}{-\sqrt{2}/2} = 1$.

Graphing Trigonometric Functions

In calculus, you will primarily need to remember how to graph just the sine and cosine functions and how to change the period and amplitude of the graph. The period of a trigonometric function is the distance along the x-axis that it takes for the graphing pattern to repeat. The amplitude is the number of units a graph rises above and sinks below the x-axis. The graphs are always done with a domain of radians, not degrees. If you are graphing just $y = \sin(x)$ or $y = \cos(x)$, then the standard period is 2π. To graph $y = a \cdot \sin(b \cdot x)$, where a and b are constants, two changes will occur. The a in front will multiply all y values on $y = \sin(x)$ by that amount. The b coefficient of x will create a new period of length $\frac{2\pi}{b}$. Often you will have a graphing calculator with which to produce the graph, but if you don't, it's good to know how to produce a basic sketch by hand. Study the following three examples to refresh your memory of the graphs of sine and cosine.

EXAMPLE 1-7

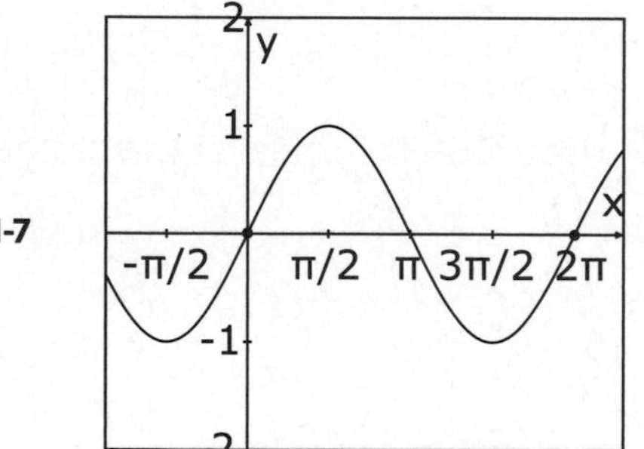

Figure 1-7

One period of the graph of $y = \sin(x)$ starts at (0, 0), has an amplitude of 1 unit, and repeats its pattern over an interval of 2π.

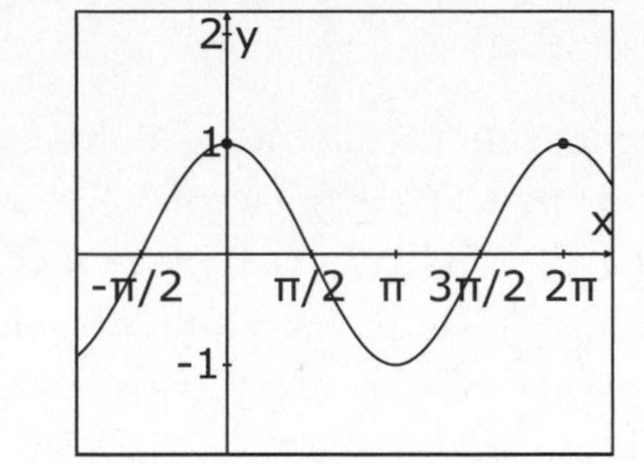

Figure 1-8

One period of the graph of $y = \cos(x)$ starts at (0,1), has an amplitude of 1 unit, and repeats its pattern over an interval of 2π.

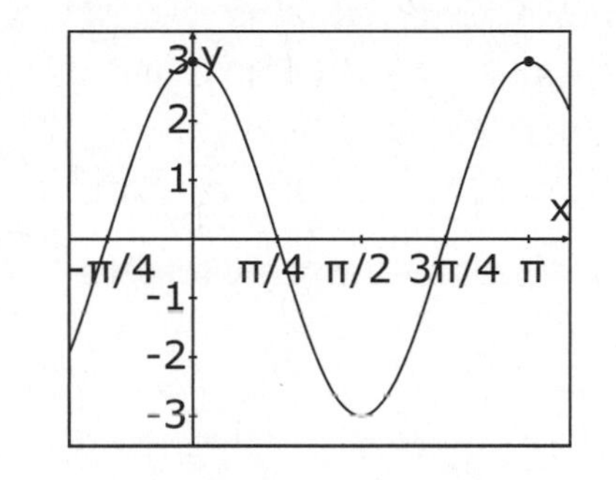

Figure 1-9

The graph of $y = 3\cos(2x)$ starts at (0,3) because of the factor of 3.

The graph has an amplitude of 3 units. It repeats its pattern over an interval of $\frac{2\pi}{2}$, or π.

Simple Trigonometric Identities

A trigonometric identity is a statement that is true for all values of the domain where the functions involved are defined. For example, $\tan(x) = \frac{\sin(x)}{\cos(x)}$ is true for all values of x, except when x is an odd multiple of $\frac{\pi}{2}$. At those locations the cosine equals 0, thereby making the tangent undefined. Even though there are a multitude of trigonometric identities, only those listed in Appendix A show up regularly in calculus.

Nine Basic Functions

Among the tools every good calculus student has at hand are all the basic graphs learned in the preceding courses. When you are faced with graphing $y=(x-3)^2+1$, you should recognize that it is just a transformed version of $y=x^2$. Replacing x with $x-3$ shifts the graph of $y=x^2$ to the right 3 units. Adding 1 shifts the graph up 1 unit. If you know the graph of $y=x^2$, then using transformations to sketch a variation is a powerful tool. The nine most commonly used graphs are summarized in Appendix A, but before looking them up, you should test yourself to see how many of the graphs you remember.

All the functions are given below. Try to sketch each one. Just concentrate on basic shape and location. After attempting a sketch, confirm (or correct) your graph by using a graphing calculator. When you can sketch each graph without assistance, check it off!

NINE COMMON GRAPHS

- ☐ $y=x$
- ☐ $y=x^2$
- ☐ $y=x^3$
- ☐ $y=|x|$
- ☐ $y=\sin(x)$
- ☐ $y=\cos(x)$
- ☐ $y=e^x$
- ☐ $y=\ln(x)$
- ☐ $y=\dfrac{1}{x}$

Functions and Composition

One of the most important skills you can bring to your study of calculus is a strong grasp of function notation and composite functions. You can break down most complicated-looking functions into a composition of simpler functions. When you encounter complicated expressions in calculus, using composite functions to tackle them in manageable steps can be very helpful.

Function notation is a shorthand method to help you work with numerous functions at once and keep them all organized. You can easily communicate the function with which to work and what to do with that function. Substituting one function into another creates a composite function.

EXAMPLE 1-8

Let $f(x)=x^2+x$ and $g(x)=x-1$.

If you are told to find $f(3)$, you know that you must work with function f and evaluate it when $x=3$.

$f(3)=3^2+3=12$

If you are told to find $f(g(6))$, evaluate g when $x=6$, and then substitute that result into function f.

$g(6)=6-1=5$

$f(g(6))=f(5)=5^2+5=30$

To find the composite function $f(g(x))$, do the same substitutions using just the variable expressions.

$$\begin{aligned} f(g(x)) &= f(x-1) \\ &= (x-1)^2+(x-1) \\ &= x^2-2x+1+x-1 \\ &= x^2-x \end{aligned}$$

Decomposing functions is a slightly more advanced but necessary skill. It requires taking a complicated expression and finding two functions whose composite is equivalent to the original function. There can be many different correct ways to decompose functions. Try to view a complicated function by seeing it as a function inside another function.

EXAMPLE 1-9

Decompose $y = \sqrt{x-7}$ into two functions.

The "outer" function is a square root function. The "inner" function is a linear function.

Let $y = \sqrt{u}$ and $u = x - 7$. Therefore, y is defined as a function of u, and u is defined as a function of x. The composite of these two functions is $y(u(x)) = \sqrt{u(x)} = \sqrt{x-1}$.

As you finish brushing up on prerequisite skills, you may find that there are certain topics you need to work on a bit more. Remember that the Internet is a great resource for math help. Because websites come and go quickly, no specific sites will be recommended here, but you can simply use a search engine and look up any of the topics in this overview. For example, doing a search for "graphing quadratic functions" will result in thousands of websites to enhance your review.

CHAPTER 2

Start Building with Limits

Almost every major concept in calculus is related to limits. It is safe to say that without them, calculus would not have developed. Understanding limits is crucial to getting off to a good start in this subject. Limits can be dealt with on a variety of levels, from intuitive to very formal. This chapter will treat them more intuitively, examining them numerically, graphically, and symbolically. You won't cover proofs with limits, however—leave that to the mathematics doctoral candidates.

Foundation of Calculus

Even though you do not need to go too deeply into limits, getting an overview of their role in the development of calculus will increase your appreciation for the incredible work done by mathematicians to establish this field of study. In formal mathematics, no new theorem can be accepted without proof. Early calculus studied mathematics on what might be termed a microscopic level, examining functions and changes on infinitely small intervals. The idea of limits was the tool devised to accomplish this task. An introductory calculus course consists of two major divisions: differential calculus and integral calculus. Both of these are established and verified using limits.

It is easy to define the slope of a line. When you look at a curved graph, though, the idea of slope becomes more challenging to grasp. But if you work with a small enough section of that curved graph, it can appear straight. You are essentially finding the limit of the slope between two points on a curve that are moving closer and closer together. Differential calculus is founded on this premise.

Why did early explorers believe the Earth was flat, even though today we know that it is relatively spherical?

Simple! In relation to the overall size of the Earth, the distances that the early explorers could view at any given time were essentially microscopically small. On a "microscopic" level, what is actually curved appeared straight.

Anybody with a bit of geometry skill can find the area inside a triangle, a trapezoid, or even a polygon of more sides. It is manageable because the objects are formed using straight lines. But what if the shape had curved sides? Then the task becomes significantly more difficult. The other major portion of the course, integral calculus, develops skills that enable us to find areas and volumes of irregularly shaped objects or surfaces. Again, limits are at the foundation of the idea; they are used to justify its validity and accuracy to the larger mathematics community. Essentially all the major concepts, and many of the smaller components of these concepts, were proved using limits. Fortunately, the forefathers of calculus have done the hardest work for you, thereby making the big ideas accessible to all who care to learn them!

Concept of a Limit

Perhaps the most important thing you should first understand about a limit is that finding a limit in math, under any circumstance, reveals where a function value, a graph, a sequence of numbers, or even a physical object is *headed*, whether or not it will ever reach that point.

Think about a segment 4 inches long that continually gets cut in half. What is the limit of the length of the remaining segment? It is clearly zero. But open your mind beyond what you can physically see. At each stage the segment has some length left, and it always will. Once it gets too small to see, you could still halve it again if you had the tools to do so, and what remained could be viewed under a microscope. And once that got too small, bring in the electron microscope! Even down to the atomic level, half of something is still something, even if it is too small for you to see or measure. The limit of the length of the remaining segment is zero inches, even though in theory it will never be completely gone.

Consider the sequence of numbers 3, 3.1, 3.01, 3.001, 3.0001 . . . . It is probably pretty clear that the numbers in the sequence are getting closer and closer to 3. Even though they will never actually reach 3, the numbers are approaching 3, and they can eventually get as close to 3 as you want; therefore, the limit is 3.

The majority of examples you will encounter in calculus deal with what is happening to a function as the independent variable approaches a number or as the absolute value of the independent variable gets infinitely large. Again, it is critical to understand that it is not necessary for the function take on the limiting value for a limit to exist. It is also important to understand that a limit reports the y value being approached as x changes. The notation looks like $\lim_{x \to 2}(3x+1)=7$. It is read, "As x approaches two, the limit of $3x+1$ equals seven." In this case, the function $3x+1$ actually takes on the limit value, seven, at $x=2$, but that is not always necessary, as you will see in later examples.

A formal definition of a limit: For a function *f* and a real number *c*, $\lim_{x \to c} f(x) = L$ if for every $\varepsilon > 0$ there exists a $\delta > 0$ such that $|x - c| < \delta \Rightarrow |f(x) - L| < \varepsilon$. This is read, "As *x* approaches *c*, the limit of *f* of *x* equals *L* if, for every epsilon greater than zero, there exists a delta greater than zero such that the absolute value of *x* minus *c* being less than delta implies that the absolute value of *f* of *x* minus *L* is less than epsilon." Whew! In much more intuitive terms, you must be able to find an interval of values around *x* that produce *y* values as close to the limit, *L*, as is required.

The formal definition of a limit is often confusing to beginning calculus students, and its use will be left for a more rigorous approach to the subject. For your purposes, focus on the intuitive idea that you should be able to find x values around a point that get you closer and closer to a given value.

One-Sided Limits

When searching for a limit at a specific x value, you can talk about three different limits. You can examine what is happening to the y value as the x value is approached from the left, you can examine what is happening to the y value as the x value is approached from the right, or you can speak of a general limit. Approaching x from the left or the right yields what is called a one-sided limit. It is distinguished from a general limit by the use of a superscript plus or minus sign after the x value in the limit notation. For example, $\lim_{x \to 3^+} x^2 = 9$ is read, "As x approaches three from the right, the limit of x squared equals nine." The plus sign after the 3 indicates that you should seek the value that y is approaching only for x values greater than 3. A negative sign after the 3 would instruct you to examine y values as x approaches 3 from the left. On a closed interval, endpoints can have only one-sided limits.

Example 2-1 uses the piecewise function $g(x)$ shown in Figure 2-1.

Figure 2-1

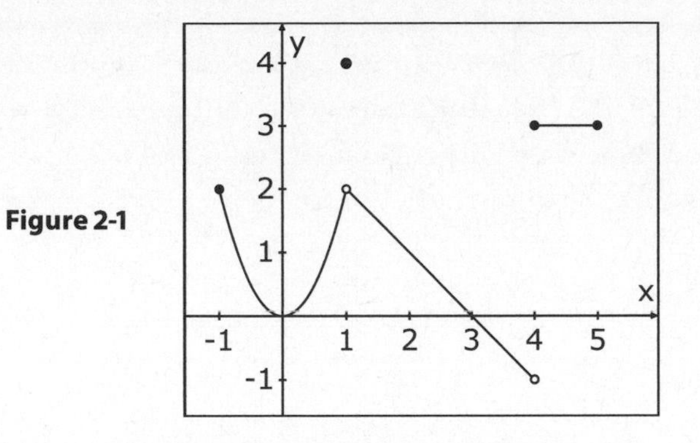

EXAMPLE 2-1

As x approaches –1 from the right, the y value approaches 2, and in this case becomes 2, so $\lim_{x \to -1^+} g(x) = 2$.

As x approaches 1 from the left, the y value approaches 2, so $\lim_{x \to 1^-} g(x) = 2$. This is true even though $g(1) = 4$, and it would still be true if $g(1)$ did not even exist.

As x approaches 4 from the left, the y value approaches –1, so $\lim_{x \to 4^-} g(x) = -1$ even though at $x = 4$, g does not equal –1. Remember that the emphasis is on what the y value is *approaching*.

As x approaches 4 from the right, the y value approaches 3, and in this case becomes 3, so $\lim_{x \to 4^+} g(x) = 3$.

Don't let the first few examples make you think all limits necessarily equal a constant. Sometimes, as x approaches a constant value, the y value can grow infinitely positive or negative without bound. Can you picture what characteristic a function might have if that were its behavior? Consider the graph of one of the nine basic functions, $y = \frac{1}{x}$. As x gets closer and closer to zero on the positive side of zero, the y values of the graph grow without bound. This is written $\lim_{x \to 0^+} \frac{1}{x} \Rightarrow \infty$. When this happens, some textbooks simply say the limit does not exist, whereas others call it infinity. As long as it is understood that infinity is not a number, but a way to indicate that the function increases without bound, either of these conventions is acceptable. The characteristic of the graph is that it has a vertical asymptote.

Limit at a Point

The general limit, or simply the "limit," of a function as x approaches a given value is defined using one-sided limits. The limit will exist if and only if the left- and right-sided limits are equal as x approaches the given value. Remember that the function value right at that point does not need to exist as long as y approaches the same value from the left and right sides. Thus, when you are trying to find a limit, unless you are sure what the function looks like, you should check the limit from the left and from the right. Most often this needs to be done with piecewise-defined functions and graphs specifically designed to test this knowledge. Graphs of equations made up of ratios involving the absolute value function can also cause a graph to behave in such a way as to yield different left- and right-sided limits at a point.

Given a function $f(x)$ and a real number c, $\lim_{x \to c} f(x) = L$ if and only if $\lim_{x \to c^+} f(x) = \lim_{x \to c^-} f(x) = L$. At a given point, if the right- and left-sided limits equal the same value L, the general limit equals L.

There are essentially four ways to determine a limit as x approaches a given value. First, you can examine the graphical behavior of a function. Second, you can carefully draw conclusions from a table. The third method is to take a numerical approach, almost creating a table of values of your own. The fourth method is an analytic approach, which usually involves simplification of the function and direct substitution of a value.

Graphical Limits

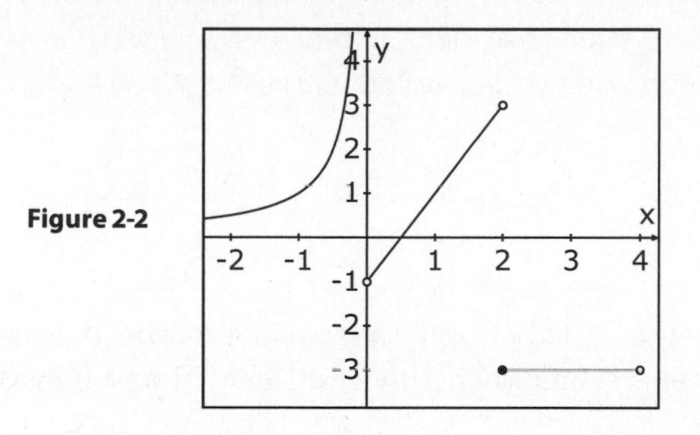

Figure 2-2

EXAMPLE 2-2

Refer to Figure 2-2, the graph of the piecewise function $f(x)$, to study and understand the following limits.

$\lim_{x \to 0^-} f(x) \Rightarrow \infty$ For $x < 0$, the function is $y = -\frac{1}{x}$, which grows without bound as x approaches zero.

$\lim_{x \to 0^+} f(x) = -1$ From the right-hand side of the y-axis, the graph approaches the point $(0,-1)$. The limit is –1 even though the function does not contain the point $(0,-1)$.

$\lim_{x \to 0} f(x)$ does not exist. The left- and right-sided limits are not equal.

$\lim_{x \to 2} f(x)$ does not exist. The limit as x approaches 2 from the right is –2 and does not equal the limit as x approaches 2 from the left, which is 3.

When a graph is provided for you in a testing situation, you can trust what you see. Any special behavior is usually described in supplemental comments. But if you produce a graph yourself using a graphing calculator, be on the lookout for hidden or incomplete behavior of the graph, which, when it occurs, is usually due to the calculator's inability to provide high enough resolution. A graph on a calculator may look like it stops at a certain point, when it actually does not stop there. In those cases, try to verify your conclusion with an additional method, such as a table of values.

Tabular Limits

You can also determine some limits from tables of values, but you must be careful not to assume too much about how a function behaves just on the basis of a few ordered pairs. The context of the problem may enable you to draw those conclusions, but beware of unsupported assumptions. Examine the ordered pairs given in Table 2-1.

TABLE 2-1

x	$h(x)$
0.99	4.8
0.999	4.99
0.9999	4.998
1.0001	5.001
1.001	5.03
1.01	5.4

Judging on the basis of the values in the table, what is a reasonable assumption for $\lim_{x \to 1} h(x)$? The way the question is posed, you can answer "5," because as x approaches 1 from the right and the left, the $h(x)$ values appear to be approaching 5. Remember, though, that you cannot tell what values $h(x)$ actually takes on, no matter how close x gets to 1. There is no certainty what the limit actually is and no certainty that it even exists, but 5 is a reasonable assumption.

There are many times when you can directly substitute into the function the value that x is approaching. However, if you try this and get anything of the form $\frac{0}{0}$, do not assume that value is 1 or 0. This is the mistake most commonly made by students working with limits. When you encounter a limit in the form of zero divided by zero, you must try a different method, such as analytically changing the form of the function or exploring it numerically.

Numerical Limits

Sometimes, when you do not have a table of values and cannot visualize what a graph looks like, a numerical approach is a valid method for determining a limit. In this situation, you substitute values for x that are very close to the number in question, in a sense creating your own table of values. If a pattern can be determined, you can find the limit or determine that one does not exist.

EXAMPLE 2-3

Find $\lim_{x \to 0} \frac{|2x|}{x}$.

You cannot just substitute 0 in for x because you would get an undefined expression, $\frac{0}{0}$.

This may also be a graph with which you are not familiar. Therefore, the next logical move is to try substituting numbers close to zero, making sure you check the left- and right-sided limits.

If $x = 0.1$, the expression becomes $\frac{0.2}{0.1}$, which equals 2.

If $x = 0.01$, the expression becomes $\frac{0.02}{0.01} = 2$. You may be thinking the limit is 2!

Now check x values to the left of zero. If $x = -0.1$, the expression becomes $\frac{0.2}{-0.1}$, or -2. The numerator stays positive because of the absolute value symbol.

If $x = -0.01$, the expression becomes $\frac{0.02}{-0.01}$, or -2.

The left-hand limit seems to be -2, and the right-hand limit seems to be 2. This is indeed the case, so the limit as x approaches 0 fails to exist. If you have a graphing calculator, use it now to get a visual reinforcement of the result.

When you are working on limits, there are definitely times when a trial-and-revise numerical method is the best choice.

Finding Limits Analytically

A lot of the functions that you have been working with so far have behaved in erratic or unusual ways. Fortunately, there are also many cases when there are few or no special circumstances. For instance, all limits involving polynomials or other smooth functions defined over all real numbers, such as $y = \sin(x)$, can be evaluated by direct substitution.

EXAMPLE 2-4

$$\begin{aligned}\lim_{x \to 5}(x^2 + 2x + 1) &= 5^2 + 2 \cdot 5 + 1 \\ &= 36\end{aligned}$$

$$\begin{aligned}\lim_{x \to \pi}[\cos(x)] &= \cos(\pi) \\ &= -1\end{aligned}$$

Rational functions, the ratio of two polynomials, do require a bit more work. You may recall from a previous course that when a factor cancels completely from the denominator of a rational function, it leaves a "hole" in the graph. Also, when in lowest form, all factors remaining in the denominator of a rational function create asymptotes at the zeros of those factors. Compare the graphs of the two rational functions shown in Figure 2-3 and Figure 2-4.

$$y=\frac{x^2-4}{x-2}$$

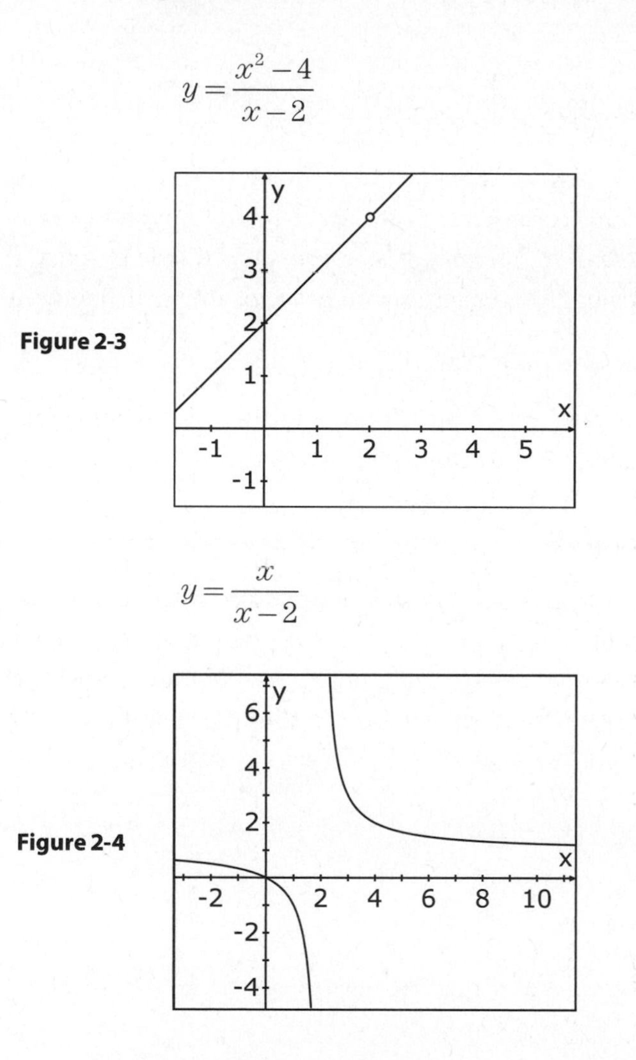

Figure 2-3

$$y=\frac{x}{x-2}$$

Figure 2-4

Note that Figure 2-3 has a "hole" in the graph at the point $(2,4)$. This is because $\frac{x^2-4}{x-2}$ reduces, and the factor $(x-2)$ cancels. The zero of $(x-2)$ is $x=2$. The rational function behaves just like its reduced form, $x+2$, everywhere except at $x=2$.

$$\frac{x^2-4}{x-2} = \frac{(x+2)(x-2)}{(x-2)}$$
$$= x+2$$

In Figure 2-4, there is a vertical asymptote at $x=2$ because the $(x-2)$ does not cancel out of the denominator. The limit as x approaches 2 does not exist.

Judging on the basis of the graph in Figure 2-3, what do you think the $\lim_{x \to 2} \frac{x^2-4}{x-2}$ equals? If you guessed 4, you are correct! This is a limit where analytic methods come to your assistance. When direct substitution gives you a form of $\frac{0}{0}$, changing the form using algebra skills is a viable option.

EXAMPLE 2-5

Find $\lim_{x \to 2} \frac{x^2-4}{x-2}$.

If you substitute $x=2$ into the expression, it becomes $\frac{0}{0}$, so you factor, reduce, and then evaluate.

$$\lim_{x \to 2} \frac{x^2-4}{x-2} = \lim_{x \to 2} \frac{(x-2)(x+2)}{x-2}$$
$$= \lim_{x \to 2}(x+2)$$
$$= 4$$

Can a limit of a function exist as *x* approaches a given value of the domain (*c*), even if the function has no *y* value at *c*?

Yes! A limit is found by finding the *y* value that the function is *approaching*. The function can have a "hole" in its graph and still have a limit at that point.

Limits As *x* Approaches Infinity

One other kind of limit used in calculus to explore the behavior of functions is a limit as the absolute value of x gets infinitely large. This limit offers insight into the behavior of the graph of the function way out on the right or left end of its domain. This is sometimes called examining end behavior, and often, the limit as x grows infinitely reveals a horizontal asymptote of a function.

There are formal and informal ways to examine a limit like this. Informally, you think about what is happening to each part of the function as x increases in magnitude and consider how fast each term is growing. Formally, you can sometimes change the form of the given function to determine a limit. A few examples will shed some light on these approaches.

As you work with limits as x approaches infinity, a simple but important limit to keep in mind is, for any positive value of p, $\lim_{x\to\infty}\left(\frac{1}{x^p}\right)=0$. Often, dividing all terms of a rational function by the highest power of x produces terms of this form.

Limits of Rational Functions

Examples 2-6 and 2-7 show two ways to find the limit of the same rational function. The first uses an analytic method. The second is an intuitive approach examining rates of growth of the numerator and denominator.

EXAMPLE 2-6

Find $\lim_{x\to\infty}\frac{2x^2-4x}{3x^2+5}$ analytically.

The highest power of x is x^2, so all terms of the expression will be divided by x^2.

$$\lim_{x\to\infty}\frac{2x^2-4x}{3x^2+5}\cdot\frac{1/x^2}{1/x^2}=\lim_{x\to\infty}\frac{2-\frac{4}{x}}{3+\frac{5}{x^2}}$$

As x gets infinitely large, $\frac{4}{x}$ and $\frac{5}{x^2}$ go to 0, so $\lim_{x\to\infty}\frac{2-\frac{4}{x}}{3+\frac{5}{x^2}}=\frac{2}{3}$.

The intuitive approach is often taught in a course leading to calculus. It simply examines the degrees of the leading terms in the numerator and denominator. As x gets very large, $2x^2$ will grow much faster than $4x$, so the value of $4x$ will become insignificant in the numerator. The same thing will happen with $3x^2$ and 5 in the denominator. As a result, for very large x values, the limit will be determined by the ratio of the leading terms.

EXAMPLE 2-7

Find $\lim_{x\to\infty}\frac{2x^2-4x}{3x^2+5}$ intuitively.

$$\lim_{x\to\infty}\frac{2x^2-4x}{3x^2+5}=\lim_{x\to\infty}\frac{2x^2}{3x^2}=\frac{2}{3}$$

The result of Examples 2-6 and 2-7 verify that the function $y=\frac{2x^2-4x}{3x^2+5}$ has a horizontal asymptote at $y=\frac{2}{3}$.

Making a Sandwich

Another method involves sandwiching a function between two simpler functions whose limits are easier to find. Frequently, this approach can be used with limits involving sines and cosines, because the values of these basic functions are limited by their amplitudes.

EXAMPLE 2-8

Find $\lim\limits_{x\to\infty} \frac{\sin(x)}{x}$.

Because $-1 \le \sin(x) \le 1$ for all real numbers, dividing all parts of the compound inequality by x will produce $\frac{-1}{x} \le \frac{\sin(x)}{x} \le \frac{1}{x}$.

Take the limit of all three parts of the inequality.

$$\lim_{x\to\infty} \frac{-1}{x} \le \lim_{x\to\infty} \frac{\sin(x)}{x} \le \lim_{x\to\infty} \frac{1}{x}$$

The ends of the inequality head to 0 and "sandwich" the middle limit.

$0 \le \lim\limits_{x\to\infty} \frac{\sin(x)}{x} \le 0$, so $\lim\limits_{x\to\infty} \frac{\sin(x)}{x} = 0$.

The limit again implies that the graph of $y = \frac{\sin(x)}{x}$ has a horizontal asymptote of $y = 0$. The graph is shown in Figure 2-5. This is an unusual case of a function that crosses its horizontal asymptote an infinite number of times!

$$y = \frac{\sin(x)}{x}$$

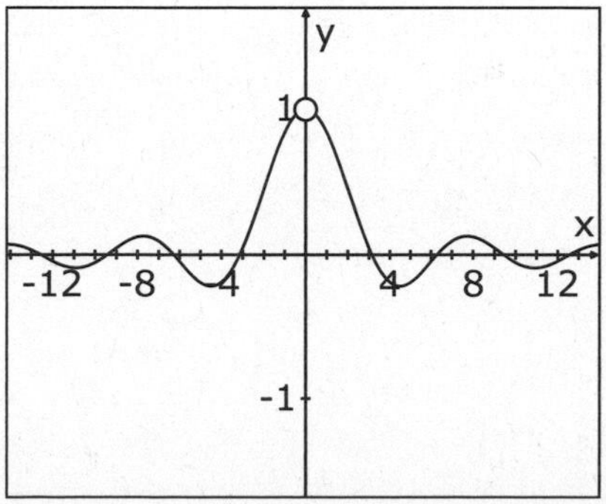

Figure 2-5

Dealing with $\frac{1}{x}$

When $\frac{1}{x}$ is part of an expression whose limit you are seeking, again it can be dealt with analytically or intuitively. The analytic approach involves substituting a new variable for $\frac{1}{x}$. The intuitive approach is simply to think about the changing value of $\frac{1}{x}$.

EXAMPLE 2-9

Find $\lim_{x \to \infty} \ln\left(\frac{1}{x}\right)$.

Because this limit involves two of the nine basic functions, it lends itself to an intuitive effort. As you develop your mathematical intuition, you'll begin to understand how certain basic functions behave. As x gets infinitely large, $\frac{1}{x}$ approaches 0 from the positive direction. This means that the input to the natural logarithm is heading toward 0. The graph of the natural logarithm has a vertical asymptote at $x = 0$ and approaches negative infinity. Thus, by this reasoning, $\lim_{x \to \infty} \ln\left(\frac{1}{x}\right) \Rightarrow -\infty$.

Using an analytic method requires replacing $\frac{1}{x}$ with a different variable. This will also change the number being approached.

Let $a = \frac{1}{x}$. As $x \to \infty$, $\frac{1}{x} \to 0^+$, so $a \to 0^+$.

$$\lim_{x \to \infty} \ln\left(\frac{1}{x}\right) = \lim_{a \to 0^+} \ln(a) \Rightarrow -\infty$$

Do you see that the substitution process simply formalizes the thinking you did intuitively?

As you think about limits in general, try to learn to recognize and categorize limit problems into the various cases discussed in this chapter. As you determine a category for the problem, try to recall the various methods that have been modeled for each kind of limit. Finally, the greater familiarity you have with the basic graphs and their behaviors, the better intuitive sense you will have to help you evaluate limits.

Skill Check

From this point on, you will find at the end of each chapter a short set of problems. These will give you an opportunity to practice the skills covered in the chapter. Try each problem by hand unless it is designated as a problem where a calculator is needed. Sometimes the calculator can be used to verify a result you found.

If you are not sure how to begin a problem, look back through the chapter to find a similar example. Solutions appear immediately after each set of Skill Check questions. Examine the solutions only after you have attempted the problems on your own.

If you feel you need more practice than the Skill Check offers, use an Internet search engine. For example, for this chapter you might search for "calculus limit practice problems." Be patient. Some sites are far better than others, so you may have to check a few sites before finding a good one. Don't give up!

In Figure 2-6, the graph of $w(x)$ is shown on the domain $[-3,3]$. Use Figure 2-6 to find the limits for numbers 1 through 6.

Figure 2-6

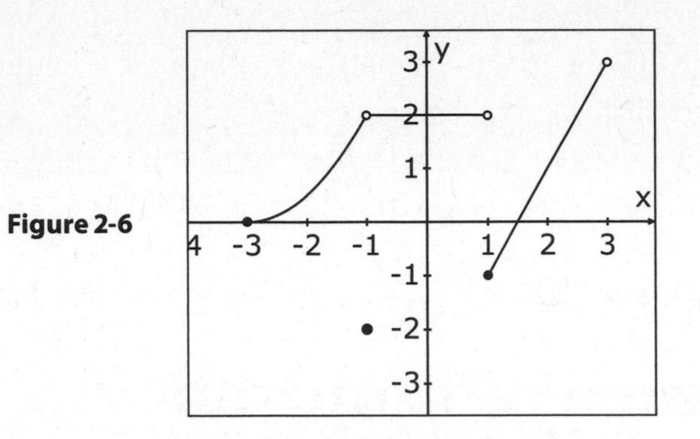

1. $\lim_{x \to -1} w(x)$

2. $\lim_{x \to 1^-} w(x)$

3. $\lim_{x \to 1^+} w(x)$

4. $\lim_{x \to 1} w(x)$

5. $\lim_{x \to 3^-} w(x)$

Find the following limits using analytic, intuitive, or numerical methods.

6. $\lim_{x \to 4} \frac{x^2 - 5x + 4}{x - 4}$

7. $\lim_{x \to \infty} \frac{7 - 5x^3}{8x^3 + 2x^2}$

8. $\lim_{x \to -4} \sqrt{12 - x}$

9. $\lim_{x \to -\infty} \cos\left(\frac{1}{x}\right)$

10. $\lim_{x \to 7^+} \frac{x+1}{x-7}$

CHAPTER 3

The Importance of Continuity

Continuity is a relatively simple but critically important concept. The hypotheses of most of the theorems in early calculus contain a reference to the necessary condition of a function being continuous. Limits are one of the cornerstones of calculus; continuity is another. Once you learn what it means, you will recognize the importance of the concept of continuity in nearly all succeeding work in calculus.

Continuity at a Point

Most people have a nonmathematical sense of what it means for something to be continuous or not to be continuous. A symphony with an intermission is not continuous, because the musicians take a break part of the way through the performance. The cycle of daylight to darkness and back to daylight is essentially continuous. In math, the same common-sense understanding of continuity can help you tell whether a function is continuous or not, but you still need to be able to approach it a bit more formally. You also need to acquire a general understanding of why continuity is important in calculus.

Simply stated, a function is continuous at a given point if there is no kind of a break in the function at that point. If this is true, then a single mathematical statement can summarize all the critical elements of this idea. A function $k(x)$ is continuous at a point $x=c$ if $\lim_{x \to c} k(x) = k(c)$. Continuity at a point is sometimes examined as consisting of three individual components.

THREE COMPONENTS OF CONTINUITY AT A POINT

1. $\lim_{x \to c} k(x)$ exists.
2. $k(c)$ exists.
3. $\lim_{x \to c} k(x) = k(c)$.

Breaking continuity at a point down into three separate parts helps you identify different ways in which continuity can fail to exist. If either of the first two conditions fails to be true, then obviously the third condition will fail as well, but it is possible for any one of the three conditions to be the primary cause of discontinuity at a point. Examine the three cases individually.

The Limit Fails to Exist

The limit can fail to exist even if the function value exists. From the earlier discussion of limits, can you picture how that might happen? If you are thinking different left-hand and right-hand limits, congratulate yourself! Figure 3-1 shows one possible scenario. Note that $k(1)$ exists, but the limit as x approaches 1 does not exist. In this figure, what is the approximate value of each one-sided limit as x approaches 1?

Figure 3-1

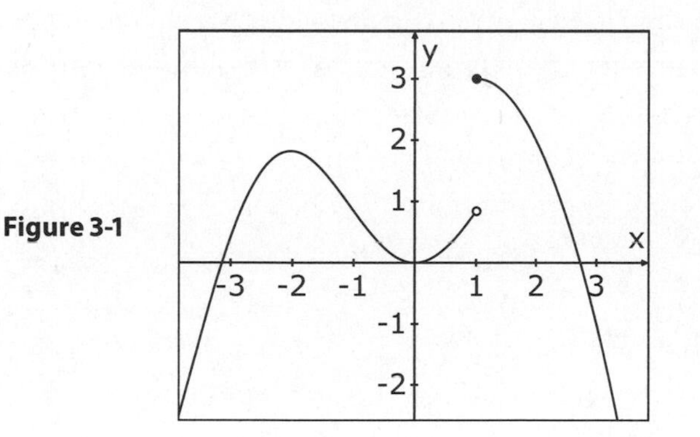

The limit approaching 1 from the left equals about 0.9. The limit approaching 1 from the right is 3. Remember that this means the general limit as x approaches 1 does not exist.

The Function Value Fails to Exist

If the limit exists, but the function value does not exist, the second component of continuity fails. This will happen with a simple "hole" in a graph, where the left- and right-hand limits are the same. If it hasn't already been stressed enough, *the limit can exist even without the function value existing*. Figure 3-2 shows this second case. The limit as x approaches 2 exists and equals –2, but $k(2)$ does not exist.

Figure 3-2

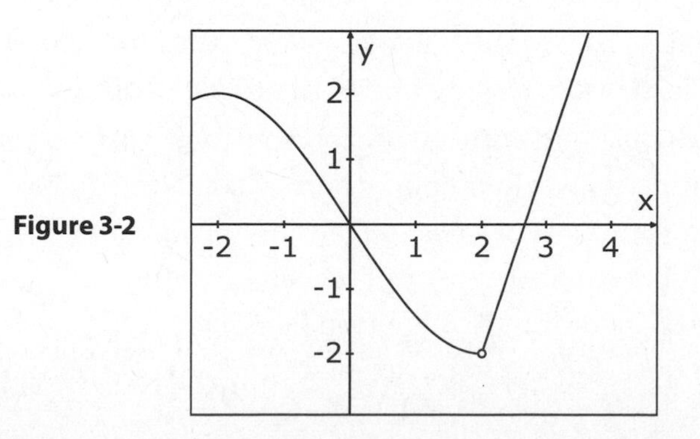

The Limit Does Not Equal the Function Value

It is also possible for the first two criteria for continuity to be satisfied and for the third component to fail. How could you change Figure 3-2 to satisfy the third component? The limit already exists. All you have to do is give $k(2)$ a function value, but make sure it is not equal to the limit. Placing a point anywhere where $x=2$, except at (2,–2), will achieve your goal.

EXAMPLE 3-1

$$\text{Let } f(x)=\begin{cases} 2x-1 & \text{for } x<0 \\ 3 & \text{for } x=0 \\ 3\sqrt{x}-1 & \text{for } x>0 \end{cases}$$

Determine whether $f(x)$ is continuous at $x=0$. If it is not continuous, explain why it fails to be continuous.

For values of $x<0$, use $2x-1$. $\lim_{x\to 0^-}(2x-1)=-1$

For values of $x>0$, use $3\sqrt{x}-1$. $\lim_{x\to 0^+}(3\sqrt{x}-1)=-1$

Because the left- and right-hand limits are equal, $\lim_{x\to 0} f(x)=-1$. But $f(0)=3$, so $f(x)$ is not continuous at $x=0$ because the limit does not equal the function value at that point.

Kinds of Discontinuity

Even though the most important skill with respect to continuity is to determine whether or not continuity exists at a point, there is some terminology that is also useful to add to your base of knowledge. The terminology gives names to the different kinds of discontinuity. The terms do not indicate which of the three conditions necessary for continuity fails; they are actually a more general way to describe the behavior of a function at a point of discontinuity. The four basic types of discontinuities that can occur at a point are a removable discontinuity, a jump discontinuity, a vertical asymptote, and an oscillating discontinuity.

Removable Discontinuity

An informal name for a removable discontinuity is a "hole" in a graph. You have already read about this kind of situation. When a factor cancels completely from the denominator of a rational function, there is a point missing from the graph at whatever value of x makes the canceled factor zero. The reason it is called removable is that you can simply assign a single function value to the graph and "fill in the hole." The way to find the function value is to plug the value of x that makes the canceled factor zero into the reduced form of the rational function. It is also possible to have a piecewise-defined function that has a hole in it at a specific point of its domain.

EXAMPLE 3-2

Redefine $h(x)=\dfrac{x^2-3x-10}{x+2}$ as a piecewise, continuous function.

Factor the numerator, and then cancel the common factor.

$$h(x)=\frac{(x-5)(x+2)}{x+2}$$

$h(x)=(x-5)$ for all values of x except $x=-2$, which is the zero of $x+2$.

The hole in the graph of h occurs at the point $(-2,y)$. The value of y is found by substituting –2 into the reduced form of $h(x)$. Thus $y=-2-5=-7$.

The piecewise function is therefore $h(x)=\begin{cases}\dfrac{x^2-3x-10}{x+2} & \text{for } x\neq-2\\ -7 & \text{for } x=-2\end{cases}$.

Jump Discontinuity

A jump discontinuity occurs when the left- and right-sided limits at a given point are not equal. In essence, the graph "jumps" from one point to another. The function value at that point can be—but does not need to be—defined. This kind of discontinuity is nonremovable, because there is no way to redefine the function to connect the points.

Figure 3-3

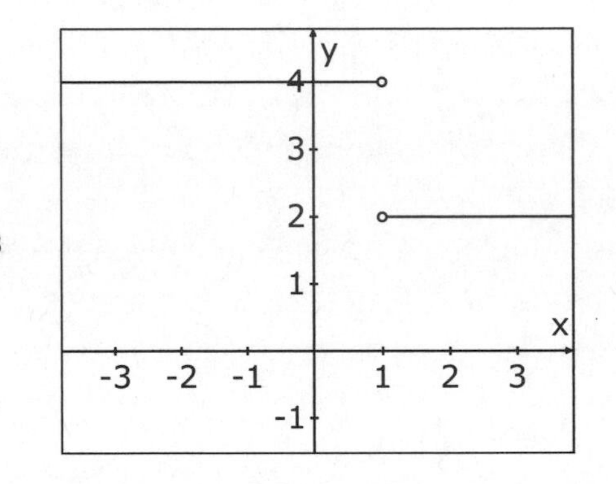

There is a jump discontinuity at $x = 1$ on the graph in Figure 3-3. The limit as x approaches 1 from the left is 4, and the limit as x approaches 1 from the right is 2. Because the y values "jump" from 4 to 2, there is no way to redefine the y value at $x = 1$ to make the graph continuous. It would make no difference whether or not a y value for $x = 1$ was defined.

Vertical Asymptote

You may have some familiarity with vertical asymptotes from a previous course, but a quick refresher never hurts. Unlike a removable discontinuity, a vertical asymptote is created when a function has a factor in its denominator that *does not* cancel. On either side of a vertical asymptote, function values approach positive or negative infinity. These discontinuities too are not removable. They are sometimes called infinite discontinuities. Example 3-3 will give you a chance to see something new and review something old.

EXAMPLE 3-3

Find all asymptotes of $p(x)=\dfrac{2x-3}{x-5}$.

The rational function does not reduce, so $x=5$ is the equation of the vertical asymptote. This is where the graph of $p(x)$ has a discontinuity.

Remember, if the limit becomes a constant as x approaches positive or negative infinity, then the function has a horizontal asymptote.

Because the numerator and denominator are of equal degree, $\lim\limits_{x\to\infty}\dfrac{2x-3}{x-5}=2$.

$p(x)$ has a horizontal asymptote of $y=2$. Horizontal asymptotes are *not* points of discontinuity.

Oscillating Discontinuity

This slightly unusual type of discontinuity is sometimes produced in trigonometric functions. The most prevalent example is $y=\sin\left(\dfrac{1}{x}\right)$ as x approaches 0. As this happens, the input to the sine function, $\dfrac{1}{x}$, grows rapidly toward positive infinity on the right of 0 and toward negative infinity on the left of 0. This rapid growth of $\dfrac{1}{x}$ causes the sine graph to oscillate wildly, as shown in Figure 3-4.

Figure 3-4

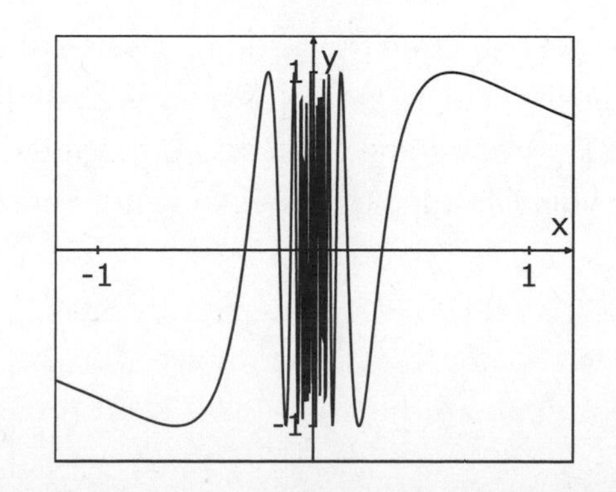

Continuity on an Interval

A function is considered continuous on an interval of its domain if it is continuous at all points in the interval. You can talk about continuity on open, closed, or half-open intervals. Any time an interval includes an endpoint, the limit in the necessary condition for continuity becomes a one-sided limit. For instance, at the right endpoint of an interval, a function is continuous if the left-sided limit is equal to the function value at that endpoint. It is understood that a right endpoint cannot be approached from the right.

Even though a lot of attention is given to continuity at a point, continuity on an interval begins to expand the information you can collect and the conclusions you can draw about the behavior of a function. Because calculus examines change over time, in order to apply many of the principles of calculus, you need to be assured that the function you are working with is continuous on the working interval. You use the idea of continuity at a point to determine continuity on an interval, which in turn enables you to apply calculus to the function on that interval. It all fits together so well! Example 3-4 reinforces the idea of continuity on an interval.

Figure 3-5

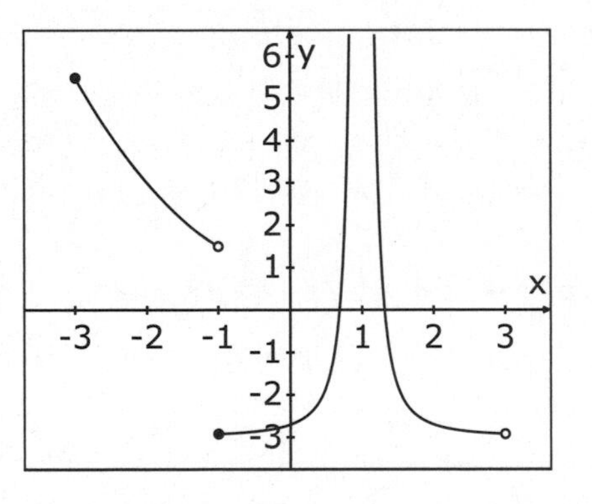

EXAMPLE 3-4

On the domain $[-3,3]$, determine all intervals of continuity for the function shown in Figure 3-5. The graph has a vertical asymptote at $x=1$.

There is a jump discontinuity at $x=-1$. There is an asymptote discontinuity at $x=1$. There is a removable discontinuity at $x=3$. Therefore, the intervals of continuity are $[-3,-1]\cup(-1,1)\cup(1,3)$.

The Intermediate Value Theorem

In addition to its presence in the hypotheses of many of the theorems of calculus, one of the first significant results of continuity was the Intermediate Value Theorem. Informally, this theorem says that if a function is continuous on a given domain, that function will take on all values between the values at each endpoint at least once in the interval. That may not seem very significant until you think about one of the key implications.

Consider a function that is continuous on a given domain. If one endpoint has a positive y value, and the other endpoint has a negative y value, there must be at least one point in the interval where the function value becomes zero. This observation enabled early mathematicians to narrow down the location of the solutions of very complicated equations, or to determine that a solution to an equation at least existed, even if they could not find the exact value of the solution. For instance, a related result was the ability to prove that every polynomial of odd degree has at least one real solution, whether or not it can actually be found.

The Intermediate Value Theorem: If $w(x)$ is continuous on the closed interval $[a,b]$, and k is any value such that $w(a) \le k \le w(b)$, then there exists at least one value c in $[a,b]$ such that $w(c) = k$. This is also true if $w(b) \le k \le w(a)$.

The Intermediate Value Theorem has also been applied to more contemporary situations. For example, when a graphing calculator finds the x-intercept of a graph, the user generally indicates a point on either side of the x-intercept. Not coincidentally, those points marked by the user have function values with opposite signs. By indicating points in this way, the user is marking a domain that satisfies the Intermediate Value Theorem, and the calculator can then find the x-intercept in that domain.

Skill Check

Try each problem by hand unless it is designated as a problem where a calculator is needed. Sometimes the calculator can be used to verify a result you found. If you are not sure how to begin a problem, look back through the chapter to find a similar example. If you feel you need more practice than the Skill Check offers, use an Internet search engine. For this chapter, you might search for “continuity practice problems.”

1. Given $h(x)=\dfrac{2x-6}{x^2+4x-21}$.

   Find the x value(s) where $h(x)$ is discontinuous, and name the type of discontinuity.

2. Given $f(x)=\dfrac{x^3-8}{x-2}$.

   Redefine f as a piecewise continuous function.

3. Given $p(x)=\begin{cases} ax^2-3 & \text{for } x\le 2 \\ 5x+1 & \text{for } x>2 \end{cases}$.

   Find a such that $p(x)$ is continuous at $x=2$.

4. 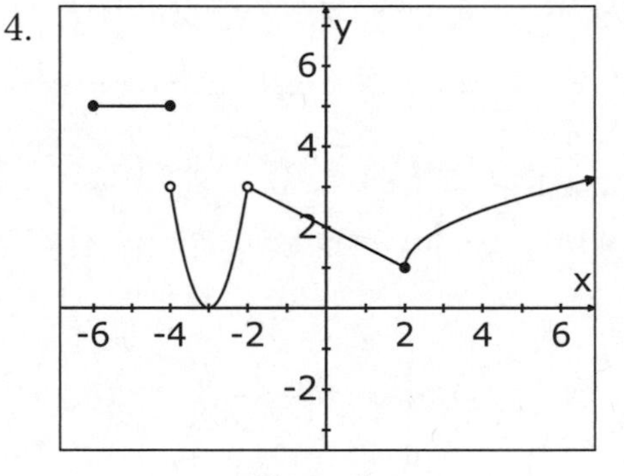

Figure 3-6

   On the domain $[-6,\infty]$, find the intervals of continuity in Figure 3-6.

5. Given $q(x)=\begin{cases} -x+2 & \text{for } x<-3 \\ x^2+x & \text{for } -3\le x<1 \\ \sqrt{x+3} & \text{for } x\ge 1 \end{cases}$.

   Find the x value(s) where $q(x)$ is discontinuous, and name the type of discontinuity.

6. Use the Intermediate Value Theorem to show that the equation $x^4-2x=3$ has at least one solution between $x=1$ and $x=2$.

CHAPTER 4

Getting Differentiability Straight

The first major portion of any calculus course is dominated by the concept of the derivative. You may not yet know exactly what derivatives are, but they are everywhere in the world around you. Velocity is a derivative. Marginal revenue is a derivative. Population growth is a derivative. Derivatives provide information about rates of change on infinitesimally short intervals, so any quantity that changes can be measured, and that change can be represented using derivatives.

Average Rate of Change

Anyone who lives close to a river is always concerned when heavy rains occur. Each riverside community has someone, or perhaps a team, to keep an eye on the river levels during those particular periods. If a river rises 3 feet above its standard level and takes 6 hours to do so, then the river has risen an average of one-half foot, or 6 inches, per hour. This is an average rate of change. This does not mean that it rose exactly one-half foot every hour. Maybe during the first hour of rain, the river only rose 2 inches. But over a larger time interval, an average was calculated. To determine the average rate of change, the only information needed was river level at the start of the rain, and river level 6 hours later. If measurements are taken more frequently in an effort to predict when the river will reach flood stage, then calculus is being used, whether it looks like it or not! Any time a quantity changes with respect to something else changing (usually time), an average rate of change can be calculated.

Average Rate of Change: Let $f(t)$ be any function. The average rate of change of $f(t)$ on the interval $[t_1, t_2]$ is calculated by applying the formula $R_{avg} = \frac{\Delta f}{\Delta t} = \frac{f(t_2)-f(t_1)}{t_2-t_1}$. The symbol Δ (the Greek capital letter delta) is a math and science shorthand notation for "change in." For example, Δf is read "change in f." The expression $\frac{\Delta f}{\Delta t} = \frac{f(t_2)-f(t_1)}{t_2-t_1}$ is called a difference quotient.

EXAMPLE 4-1

Find the average rate of change of the function $h(x) = x^2 - 4$ on the interval $[-1,3]$.

Using the formula for average rate of change yields

$$\begin{aligned}R_{avg} &= \frac{h(3)-h(-1)}{3-1}\\ &= \frac{(3^2-4)-[(-1)^2-4]}{3-(-1)}\\ &= \frac{5-(-3)}{4}\\ &= 2\end{aligned}$$

Geometrically, this problem has calculated the slope of a secant on the graph of $h(x)$, as shown in Figure 4-1.

Figure 4-1

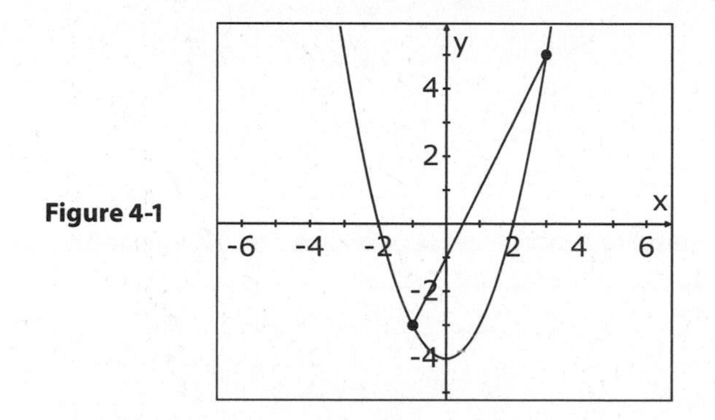

Example 4-1 does not have any units associated with it, but nearly all application problems include units to give the problem relevance. The most common independent variable is time, but time is not always the independent variable, as evidenced by Example 4-2.

EXAMPLE 4-2

Scuba divers must be careful about their rate of ascent from deep water, because an abrupt reduction in pressure can cause the nitrogen that had, in the deep water, gone into solution within the body to be released too rapidly as bubbles, resulting in great pain and often endangering the diver's life. At a depth of 20 meters (m), the water pressure is approximately 43.9 pounds per square inch (psi). At a depth of 45 meters, the water pressure is approximately 80.3 psi. Calculate the average rate of change in pressure as a diver ascends.

Because the diver is below the surface of the water, consider the depths as negative numbers.

Keep track of units as you calculate the change.

$$\frac{80.3-43.9\,\text{psi}}{-45-(-20)\,\text{m}} \approx -1.5\,{}^{\text{psi}}\!/_{\text{m}}$$

This means that as a diver ascends, the water pressure decreases about 1.5 pounds per square inch per meter. Knowing this enables divers to calculate a safe rate at which to ascend to the surface.

You always calculate an average rate of change over an interval of time and use subtraction to do so. The word *average* may make you think you must add, but the real emphasis is on change. Change is measured by finding differences. Remember that average rate of change is simply slope, and slope is the difference of the dependent variables divided by the difference of the independent variables.

If the formula for average rate of change looks familiar to you, that is not surprising. It is actually just the formula for calculating slope of a line. Average rate of change is really just an algebra concept with attention to units! One reason why it is important is that it is a doorway to the calculus concept of instantaneous rate of change.

Instantaneous Rate of Change

When you take a long car trip, your average rate is simple to find. Just take the distance traveled divided by the time it took to cover those miles. For instance, if you covered 400 miles in 8 hours, your average speed was 400 miles divided by 8 hours, or 50 miles per hour. Clearly, you were not moving at 50 miles per hour at all times. Rather, 50 mph is just an average rate of change in position over time.

In calculus terms, your speed at any moment, which is shown on your speedometer, is called your instantaneous rate of change. But say your speedometer is broken. Is there any way to estimate your speed, or instantaneous rate of change? All major highways have mile markers, so one way might be to time how long it takes you to travel between consecutive mile markers and do a little converting to find miles per hour. But if you think about it, that is still only an average rate of change over a shorter distance.

To get a more accurate measure of your speed at any moment, you would need to be able to shorten the time interval over which you measured your distance. For instance, if you had a way to know that you covered 88 feet in 1 second, the conversion would tell you that your speed was 60 miles per hour. Are you getting the idea? Instantaneous rate of change is found by calculating the average rate of change over the shortest possible interval as the change in the independent variable heads toward zero. Fortunately, limits enable you to calculate that rate.

Instantaneous Rate of Change: Let $f(t)$ be any function. The instantaneous rate of change of $f(t)$ at any given moment is calculated by $R_{\text{inst}} = \lim_{\Delta t \to 0} \frac{\Delta f}{\Delta t}$. An important difference is that an average rate of change is calculated over an interval, whereas an instantaneous rate of change is calculated at a point in time.

Revisiting Example 4-1 in two different ways should help clarify what you are finding and how to actually find it. You have already calculated the average rate of change over the interval [–1,3] to be 2, but what if you wanted to know an instantaneous rate of change at $x = 2$?

EXAMPLE 4-3

Estimate the instantaneous rate of change of $h(x) = x^2 - 4$ at $x = 2$ by finding the average rate of change of two points that are very close together.

$$\begin{aligned} R_{\text{avg}} &= \frac{h(2.01) - h(2)}{2.01 - 2} \\ &= \frac{(2.01^2 - 4) - (2^2 - 4)}{0.01} \\ &= \frac{0.0401}{0.01} \\ &= 4.01 \end{aligned}$$

This is a pretty good estimate, but it could be improved by using points even closer together. Try it yourself with $x = 2$ and $x = 2.005$. The result you should get is 4.005.

EXAMPLE 4-4

Use limits to determine the actual instantaneous rate of change of $h(x) = x^2 - 4$ at $x = 2$.

Figure 4-2 shows a close-up of the graph of $h(x) = x^2 - 4$ near $x = 2$. Choose a point close to (2, 0) by moving a small distance either side of $x = 2$. This small distance is often called Δx or h, as labeled in Figure 4-2. Now calculate the limit of the average rate of change, the slope between points P and Q on $h(x)$, as Δx gets smaller and smaller.

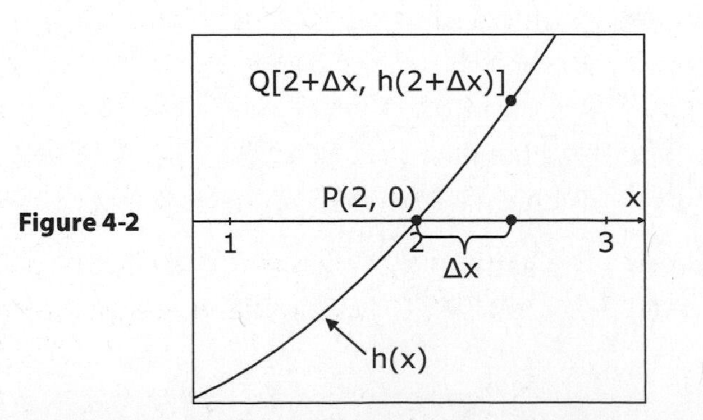

Figure 4-2

$$R_{\text{inst}} = \lim_{\Delta x \to 0} \frac{h(2+\Delta x) - h(2)}{(2+\Delta x) - 2}$$ Set up the limit of the average rate of change.

$$R_{\text{inst}} = \lim_{\Delta x \to 0} \frac{[(2+\Delta x)^2 - 4] - [2^2 - 4]}{\Delta x}$$ Substitute expressions into function h.

$$R_{\text{inst}} = \lim_{\Delta x \to 0} \frac{[4 + 4\Delta x + (\Delta x)^2 - 4] - 0}{\Delta x}$$ Expand the binomial squared using algebra.

$$R_{\text{inst}} = \lim_{\Delta x \to 0} \frac{4\Delta x + (\Delta x)^2}{\Delta x}$$ Simplify the numerator.

$$R_{\text{inst}} = \lim_{\Delta x \to 0} \frac{\Delta x(4 + \Delta x)}{\Delta x}$$ Factor the numerator.

$$R_{\text{inst}} = \lim_{\Delta x \to 0} (4 + \Delta x) = 4$$ Cancel and evaluate the limit.

Your estimate in Example 4-3 probably gave you a sense that the instantaneous rate of change was 4, but now you have just seen how calculus guarantees that it actually *is* 4. The next important hurdle beyond the algebraic manipulation is to understand fully what you have calculated. The best way to do this is by taking a graphical and geometric look at what just happened.

Tangent Lines

When the points P and Q on the graph of the function are "far" apart, it is clear that the line connecting them is a secant line. But as Δx gets smaller and smaller, the line begins to look more like a tangent line. In fact, when a tangent line exists at a point on a graph, as Δx goes toward zero, the limit of the slope of the secant line is the slope of the tangent line. The graph is said to be locally linear. Figure 4-3 is similar to a drawing seen in almost every beginning calculus book. It shows the progression of secant lines approaching a tangent line as Δx gets smaller and point Q approaches point P.

Figure 4-3

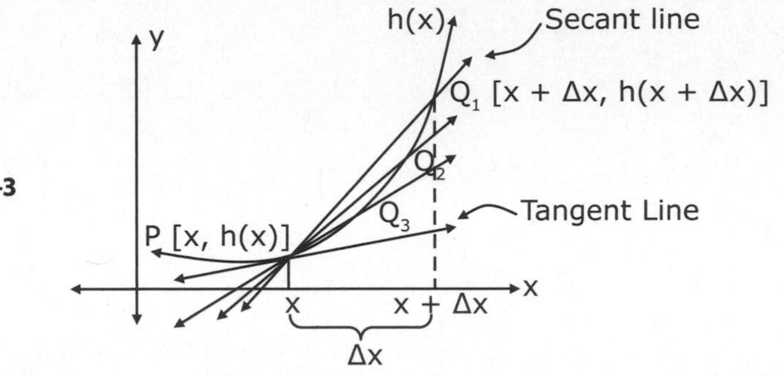

A static drawing does leave something to be desired. Fortunately, there are plenty of animations of this phenomenon on the Internet. If you are having trouble picturing the process, use a search engine and search for "secant to tangent animation." You will find that plenty of good websites show this transformation dynamically.

The important thing to realize is that through the use of limits, you can find the slope of a tangent line to a curved function at a given point, and that the slope of the tangent line is the instantaneous rate of change of the function. If a graph has local linearity at a point, then the tangent line is defined to be the slope of the function at that point. These are foundational ideas that you should try to grasp before moving forward. Study the next two examples carefully.

EXAMPLE 4-5

Find the equation of the line tangent to the function $f(x)=x^3+2$ at $x=1$.

You need the ordered pair and the slope when $x=1$. Finding the ordered pair is the easier task.

Because $f(1)=1^3+2=3$, the ordered pair is $(1,3)$.

The slope of the tangent is the limit of the average rate of change between $(1,3)$ and $[1+\Delta x, f(1+\Delta x)]$ as Δx approaches zero.

$$m_{\tan} = \lim_{\Delta x \to 0} \frac{f(1+\Delta x) - f(1)}{(1+\Delta x) - 1}$$

$$m_{\tan} = \lim_{\Delta x \to 0} \frac{[(1+\Delta x)^3 + 2] - 3}{\Delta x}$$ Substitute into $f(x)$.

$$m_{\tan} = \lim_{\Delta x \to 0} \frac{1 + 3\Delta x + 3(\Delta x)^2 + (\Delta x)^3 + 2 - 3}{\Delta x}$$ Expand the binomial cubed.

$$m_{\tan} = \lim_{\Delta x \to 0} \frac{3\Delta x + 3(\Delta x)^2 + (\Delta x)^3}{\Delta x}$$ Cancel the constants.

$$m_{\tan} = \lim_{\Delta x \to 0} \frac{\Delta x[3 + 3\Delta x + (\Delta x)^2]}{\Delta x}$$ Factor Δx out of the numerator.

$$m_{\tan} = \lim_{\Delta x \to 0} [3 + 3\Delta x + (\Delta x)^2] = 3$$ Reduce and evaluate.

$$y - 3 = 3(x - 1)$$ Write the equation of the tangent line.

EXAMPLE 4-6

A toy car is moving along a straight track. Let $s(t) = \sqrt{t}$ be the distance, in feet, of the car from its starting point at time t seconds after starting. How fast is the car moving when $t = 4$ seconds?

The speed of the car at an instant in time is the limit of the average rate of change as Δt approaches zero. Use $(4,2)$ and $[4+\Delta t, s(4+\Delta t)]$.

$$\text{Velocity} = \lim_{\Delta t \to 0} \frac{s(4+\Delta t) - s(4)}{(4+\Delta t) - 4}$$

$$\text{Velocity} = \lim_{\Delta t \to 0} \frac{\sqrt{4+\Delta t} - 2}{\Delta t}$$ Substitute into $s(t)$.

$$\text{Velocity} = \lim_{\Delta t \to 0} \frac{\sqrt{4+\Delta t} - 2}{\Delta t} \cdot \frac{\sqrt{4+\Delta t} + 2}{\sqrt{4+\Delta t} + 2}$$

Rationalize the numerator. A new skill!

$$\text{Velocity} = \lim_{\Delta t \to 0} \frac{4+\Delta t - 4}{\Delta t(\sqrt{4+\Delta t} + 2)}$$

Simplify the numerator.

$$\text{Velocity} = \lim_{\Delta t \to 0} \frac{1}{\sqrt{4+\Delta t} + 2}$$

Cancel the constants and Δt.

$$\text{Velocity} = \frac{1}{\sqrt{4+0} + 2} = \frac{1}{4} \frac{\text{ft}}{\text{sec}}$$

Evaluate the limit.

Definition of the Derivative

Because the instantaneous rate of change occurs so regularly, the result is given a name. It is called a derivative. What you have already been doing in this chapter is finding derivatives! In applications it has many interpretations, but in simplest terms, the derivative is the instantaneous rate of change of a quantity. Thus, if you are asked to find a derivative of a function at a given point, you simply use limits to find the instantaneous rate of change of the function. Remember, another common interpretation is that you are finding the slope of the line tangent to the graph of the function, which is also considered the slope of the function.

Definition of the Derivative (Difference Quotient Form): Let $y = f(x)$ be any function continuous on some interval containing a point $[c, f(c)]$. If it can be calculated, the derivative of $y = f(x)$ at a point $x = c$ is $\lim_{\Delta x \to 0} \frac{\Delta y}{\Delta x} = \lim_{\Delta x \to 0} \frac{f(c+\Delta x) - f(c)}{\Delta x}$. If the derivative exists, then f is said to be *differentiable.* The derivative of $f(x)$ at $x = c$ is the slope of f at c.

For simplicity of notation, from this point on, we will use h instead of Δx. Remember that it still stands for a small change in x.

EXAMPLE 4-7

Find the derivative of $f(x) = -x^2 + 5$ at (1,4).

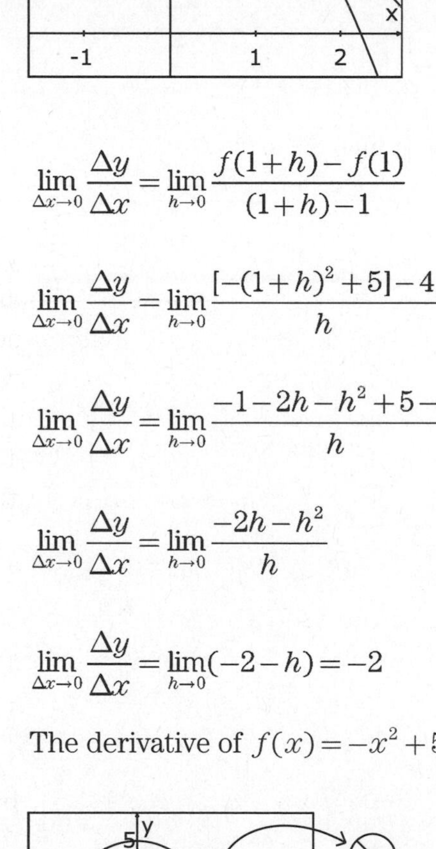

Figure 4-4

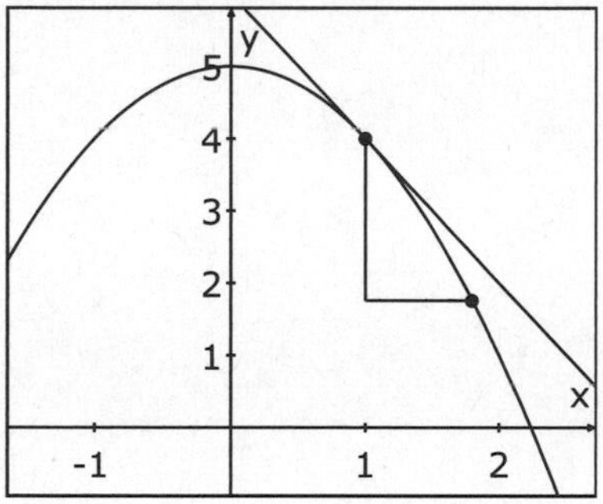

$$\lim_{\Delta x \to 0} \frac{\Delta y}{\Delta x} = \lim_{h \to 0} \frac{f(1+h) - f(1)}{(1+h) - 1}$$

$$\lim_{\Delta x \to 0} \frac{\Delta y}{\Delta x} = \lim_{h \to 0} \frac{[-(1+h)^2 + 5] - 4}{h}$$

$$\lim_{\Delta x \to 0} \frac{\Delta y}{\Delta x} = \lim_{h \to 0} \frac{-1 - 2h - h^2 + 5 - 4}{h}$$

$$\lim_{\Delta x \to 0} \frac{\Delta y}{\Delta x} = \lim_{h \to 0} \frac{-2h - h^2}{h}$$

$$\lim_{\Delta x \to 0} \frac{\Delta y}{\Delta x} = \lim_{h \to 0}(-2 - h) = -2$$

The derivative of $f(x) = -x^2 + 5$ at (1,4) is –2.

Figure 4-5

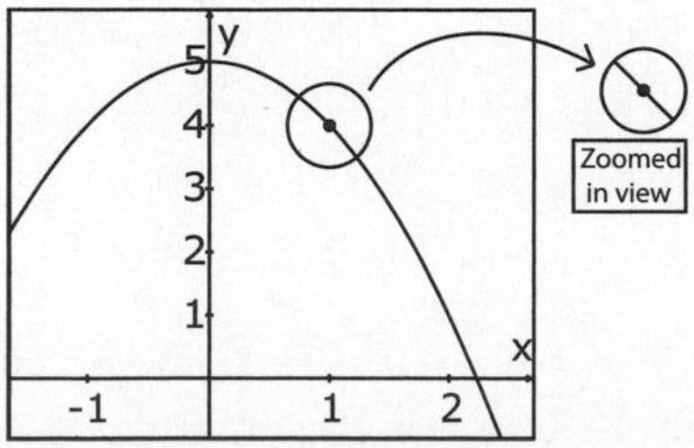

Figure 4-5 shows how the graph of $f(x) = -x^2 + 5$ looks locally linear as you take a "microscopic" look at the graph near (1,4). You could achieve this yourself by zooming in around (1,4) on a graphing calculator.

Conditions for Differentiability

Note that the definition of the derivative includes the phrase "if it can be calculated." There are certain necessary conditions for a derivative to exist at a given point. Examine the expression for the derivative a little more closely.

$$\lim_{\Delta x \to 0} \frac{\Delta y}{\Delta x} = \lim_{\Delta x \to 0} \frac{f(c + \Delta x) - f(c)}{\Delta x}$$

A derivative is a limit, so in order for the derivative to exist, the limit must exist. Thinking back to what you learned about limits, recall that a limit will fail to exist if it the one-sided limits are different or if the value of the limit heads toward positive or negative infinity.

The other way the derivative at a point will fail to exist is if the function is not continuous at that point. In the limit expression for the derivative, the actual function value must exist.

Ways the Derivative Fails to Exist

To reinforce the conditions for differentiability, it helps to have a visual understanding of the conditions under which a derivative will not exist. There are four ways a derivative will fail to exist at a point [c,$f(c)$].

- The graph of the function is not continuous at $x = c$. The limit cannot be calculated without $f(c)$.
- The graph has a vertical tangent at $x = c$. The limit heads toward positive or negative infinity.
- The graph has a corner at $x = c$. The one-sided limits will be unequal.
- The graph has a cusp at $x = c$. A cusp is a special case of a corner. The slopes of the curve on either side of the point approach positive and negative infinity.

Example 4-8 examines the derivative limit from each side and shows more specifically why a derivative does not exist at a corner. The detailed work to find each limit is not shown, because the result is what we want to focus on.

EXAMPLE 4-8

Given $g(x)=\begin{cases}-3x+4 & \text{for } x<1\\ 0.5x^2+0.5 & \text{for } x\geq 1\end{cases}$.

Use limits to show that the derivative does not exist at $x=1$.

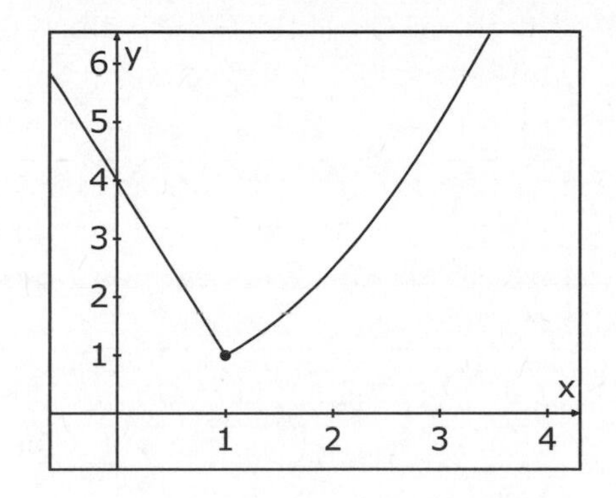

Figure 4-6

The left-sided limit finds the slope using $-3x+4$. The slope of the line is -3.

$$\begin{aligned}\lim_{h\to 0^-} g(x) &= \lim_{h\to 0^-}\frac{[-3(1+h)+4]-[-3\cdot 1+4]}{h}\\ &= -3\end{aligned}$$

The right-sided limit finds the slope using $0.5x^2+0.5$.

$$\begin{aligned}\lim_{h\to 0^+} g(x) &= \lim_{h\to 0^+}\frac{[0.5(1+h)^2+0.5]-[0.5(1)^2+0.5]}{h}\\ &= 1\end{aligned}$$

Because the slopes are different from each side, the limit—and therefore the derivative—does not exist at $x=1$, even though the function is continuous at $x=1$.

Skill Check

For this Skill Check, try each problem by hand. Sometimes the calculator can be used to verify a result you found. If you are not sure of how to begin a problem, look back through the chapter to find a similar example.

1. A pool is being drained for the winter. At 10 A.M., there are 42,500 gallons in the pool. At 2 P.M., there are 34,000 gallons in the pool. Find the average rate of change of gallons over this time interval.
2. Find the average rate of change of $p(x)=\frac{1}{x}$ on the interval [2,6].
3. The following table shows the increasing speed of a car, in feet per second, measured every 2 seconds over an interval of 8 seconds. Given that acceleration is the instantaneous rate of change of velocity over time, use the table values to provide a best estimate of the acceleration at $t=5$ seconds.

Time (seconds)	0	2	4	6	8
Speed (feet per second)	0	28	47	62	74

4. Ignoring air resistance, the distance a freely falling object has dropped is modeled by $d(t)=4.9t^2$, where time is in seconds and distance is in meters. How fast is an object falling 2 seconds after it is dropped?
5. Find the equation of the line tangent to $f(x)=x^3-4$ at $x=-1$.
6. Why does the derivative of $y=\frac{3}{x-7}$ fail to exist at $x=7$?
7. Why does the derivative of $y=|x+3|$ fail to exist at $x=-3$?

CHAPTER 5

Derivatives of Polynomials

Once you grasp the idea of a derivative as a limit, the next goal is to begin to apply that understanding to the functions you work with frequently. The functions that appear most commonly in early mathematics studies are polynomials. You began to work with derivatives of polynomials in the previous chapter, but there is much more to know. This chapter will extend those ideas and introduce more efficient ways to find derivatives and deal with more complex situations.

A Second Definition of the Derivative

With just a slight change of notation, the definition of the derivative at a point, from the previous chapter, can be made more efficient. This refined definition reduces the amount of algebraic manipulation and makes it possible to arrive at answers more quickly. The second definition is also used when you want a derivative at a known point.

In the first definition, the difference quotient form, the limit used the horizontal distance between two points approaching zero. The change in notation for the second definition, sometimes called the "x approaches a" form, comes from examining the limit based on a general x value approaching a known x value. The following analogy may help. The difference between the two definitions is analogous to traveling from Chicago to New York and either saying, "My distance from New York is becoming zero miles" or saying, "I am heading toward New York." They both mean the same thing.

Definition of the Derivative ("x Approaches a" Form): Let f be a continuous function on an interval containing a known point $[a, f(a)]$, and let $[x, f(x)]$ be any general point on the graph of the function. The derivative of f at a is $\lim_{x \to a} \frac{f(x)-f(a)}{x-a}$.

Figure 5-1

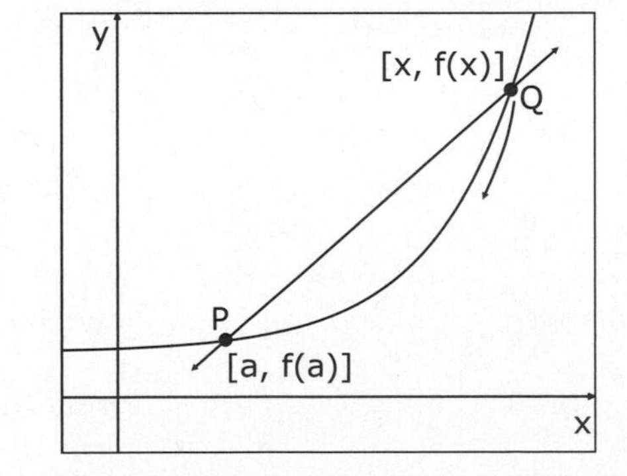

As x approaches a, the horizontal distance between points P and Q, Δx in the difference quotient definition, approaches zero. Also, the slope of the secant line becomes the slope of the tangent line. Example 5-1 shows the efficiency of this method for finding a derivative value.

EXAMPLE 5-1

Find the derivative of $r(x)=3x^2-7$ at the point $(2,5)$.

The known x value is 2, so that is a in the formula.

$$\lim_{x\to 2}\frac{r(x)-r(2)}{x-2}=\lim_{x\to 2}\frac{(3x^2-7)-5}{x-2}$$ Substitute into $r(x)$ and $r(2)$.

$$\lim_{x\to 2}\frac{(3x^2-7)-5}{x-2}=\lim_{x\to 2}\frac{3x^2-12}{x-2}$$ Simplify the numerator.

$$\lim_{x\to 2}\frac{3x^2-12}{x-2}=\lim_{x\to 2}\frac{3(x-2)(x+2)}{x-2}$$ Factor the numerator.

$$\lim_{x\to 2}\frac{3(x-2)(x+2)}{x-2}=\lim_{x\to 2}[3(x+2)]=12$$ Cancel $(x-2)$ and evaluate the limit.

Standard Notations

One of the minor challenges in learning calculus is getting used to the notation that is used. You will be more comfortable if you become familiar with the variety of different ways to denote a derivative. Some of the more common notations follow. Each is accompanied by a brief explanation of its context, its benefits, or its drawbacks.

$\frac{dy}{dx}$ is read "dy dx." The differentials dy and dx show that you have taken a limit of $\frac{\Delta y}{\Delta x}$ as Δx approached zero. You are differentiating y with respect to x. This notation identifies the independent and dependent variables. If a function had a dependent variable of p and an independent variable of t, then you would write $\frac{dp}{dt}$.

$f'(x)$ is read "f prime of x." This notation is frequently used in conjunction with information given in function notation. It identifies the function, identifies the independent variable, and can be used to show where a derivative value was found. For instance, the result of Example 5-1 can be easily summarized by $r'(2) = 12$. You found the derivative of function r to be 12 at $x = 2$.

f' is simply read "f prime." If the function were given as $w(x)$, then you could write w'. This function is used when the original information is in function notation. It is a little less formal and does not identify the independent variable.

y' is read "y prime" and can be used when a function is given in "y =" form. This notation too is very informal and does not identify an independent variable.

There is one more notation that is a little different from the previous four. $\frac{d}{dx}$ (any function of x) is read "d dx" of whatever follows it. What makes this different is that the $\frac{d}{dx}$ is being used as an operator on what follows it. For instance, $\frac{d}{dx}(x^3 + 7x)$ tells the reader to take the derivative of $(x^3 + 7x)$ with respect to x.

Even though this may seem to be a lot of notation, all of these standard notations mean the same thing, and being comfortable with each will make studying calculus a lot easier.

Derivative at Any Point

So far, the emphasis has been on finding the value of the derivative at a known point, a fixed value of x. Taking all those limits would be very tedious if you wanted the slope of a function at many points. Fortunately, in mathematics the next logical step beyond considering specific cases is to generalize concepts. This helps streamline many tasks. For example, after students have learned the long division of polynomials, the more efficient synthetic division is introduced.

So what will it take to generalize the idea of a derivative of a function at any point? Simply work the limit algorithm with a variable in the expression, rather than a constant. Example 5-2 introduces this idea. The difference quotient form of the derivative is actually easier to use in this case. The variable x will be used where a constant has been used in all previous examples.

EXAMPLE 5-2

Find the slope of $g(x)=x^2$ at any point on the graph.

$$g'(x)=\lim_{h\to 0}\frac{f(x+h)-f(x)}{(x+h)-x}$$ Set up the difference quotient.

$$g'(x)=\lim_{h\to 0}\frac{(x+h)^2-x^2}{h}$$ Substitute for the function notation.

$$g'(x)=\lim_{h\to 0}\frac{x^2+2xh+h^2-x^2}{h}$$ Expand the binomial squared.

$$g'(x)=\lim_{h\to 0}\frac{h(2x+h)}{h}$$ Combine terms and factor.

$$g'(x)=\lim_{h\to 0}(2x+h)=2x$$ Reduce and substitute $h=0$.

Instead of being a numerical answer, the $2x$ result means that at any x value on the graph of $y=x^2$, the slope of the tangent line is always twice the x value. Now if you want to know the slope of the curve on $y=x^2$ at $x=-3$, you know that $g'(-3)=2(-3)=-6$.

Note that nearly all of the examples for finding derivatives, whether at specific values or a generalized x, have used fairly simple polynomials. That is because they are the easiest to work with and are a good point of entry into the idea of derivatives. There are a multitude of other functions that are much harder to deal with, so you will turn to them a bit later. Example 5-3 shows that working on a simple rational function can get more involved rather quickly.

EXAMPLE 5-3

Find a derivative formula for $r(x)=\frac{1}{x}$.

$$r'(x)=\lim_{h\to 0}\frac{r(x+h)-r(x)}{h}$$

Set up the difference quotient.

$$r'(x)=\lim_{h\to 0}\frac{\frac{1}{x+h}-\frac{1}{x}}{h}$$

Substitute for the function notation.

$$r'(x)=\lim_{h\to 0}\frac{\frac{1}{x+h}-\frac{1}{x}}{h}\cdot\frac{x(x+h)}{x(x+h)}$$

Multiply by the LCD to clear the denominators.

$$r'(x)=\lim_{h\to 0}\frac{x-(x+h)}{hx(x+h)}$$

Distribute the LCD and cancel appropriately.

$$r'(x)=\lim_{h\to 0}\frac{-h}{hx(x+h)}$$

Combine like terms in the numerator.

$$r'(x)=\lim_{h\to 0}\frac{-1}{x(x+h)}=\frac{-1}{x^2}$$

Cancel h and substitute 0 for the remaining h.

You can see that the algebra applied to simplify the difference quotient was somewhat more complicated, but it led to a nice generalization of the slope of $y=\frac{1}{x}$ at any x value on its graph.

There are two other derivatives that are fairly simple to deduce from previous discussions. Because derivatives find slopes of functions, the derivative at any point on a linear function is just the slope of the line. Also, the derivative anywhere on a constant function is always zero, because it has a slope of zero.

Derivative Rules for Monomials and Polynomials

Limits, limits, and more limits! Are you tired of limits yet? Well, except for justifying a few more results, limits are about to take a back seat to the results they enable you to develop. The previous two examples just determined the derivatives of two different powers of x. The first showed that $\frac{d}{dx}(x^2)=2x$. The next showed that $\frac{d}{dx}\left(\frac{1}{x}\right)=\frac{-1}{x^2}$. But they both still required the use of limits. Wouldn't it be nice to know a formula for the derivative of any power of x? Limits and good algebra skills will enable you to determine the formula. Try not to get bogged down in the steps of Example 5-4. The final result is the tool you will use regularly throughout your study of calculus.

EXAMPLE5-4

Find the derivative of a power of x, $f(x)=x^n$, where n is any whole number.

$$\frac{d}{dx}(x^n)=\lim_{h\to 0}\frac{f(x+h)-f(x)}{h}$$ Set up the difference quotient.

$$\frac{d}{dx}(x^n)=\lim_{h\to 0}\frac{(x+h)^n-x^n}{h}$$ Substitute for the function notation.

The hardest part is to expand $(x+h)^n$. Most preparatory courses contain the binomial expansion pattern that is used in the next step. It is generally studied in conjunction with Pascal's Triangle.

$$\frac{d}{dx}(x^n)=\lim_{h\to 0}\frac{x^n+nx^{n-1}h+\binom{n}{2}x^{n-2}h^2+\cdots+nxh^{n-1}+h^n-x^n}{h}$$

$$\frac{d}{dx}(x^n) = \lim_{h \to 0} \frac{nx^{n-1}h + \binom{n}{2}x^{n-2}h^2 + \cdots + nxh^{n-1} + h^n}{h}$$

Cancel the x^n terms.

$$\frac{d}{dx}(x^n) = \lim_{h \to 0} \frac{h\left[nx^{n-1} + \binom{n}{2}x^{n-2}h^1 + \cdots + nxh^{n-2} + h^{n-1}\right]}{h}$$

Factor h out of each term.

$$\frac{d}{dx}(x^n) = \lim_{h \to 0}\left[nx^{n-1} + \binom{n}{2}x^{n-2}h^1 + \cdots + nxh^{n-2} + h^{n-1}\right] \qquad \text{Cancel } h.$$

All terms except the first have h in them and drop out when h goes to zero.

$$\frac{d}{dx}(x^n) = nx^{n-1}$$

Without going into greater detail, this rule will work not just for whole-number powers of x but for all real-number powers.

The Derivative of a Monomial: The derivative of a single power of x is $\frac{d}{dx}(x^n) = n \cdot x^{n-1}$, where n is any real number. The derivative has a simple pattern. Bring the power down in front of x and reduce the power by 1. If there is a constant multiplier in front of the x before you begin, it just sits there and gets multiplied by n. Thus, if a is a constant, $\frac{d}{dx}(a \cdot x^n) = n \cdot a \cdot x^{n-1}$.

Now that you know the rule, you can apply it to any simple power of x without having to use limits! Look back at the derivative of x^2. The pattern says that you should bring the 2 down in front and reduce the exponent by 1, producing $2x^{2-1} = 2x$.

The derivative of $\frac{1}{x}$ came out to $\frac{-1}{x^2}$. How did that happen? Remember that $\frac{1}{x}$ can also be written as x^{-1}. If you bring the –1 down in front and reduce the exponent by 1, you get $\frac{d}{dx}(x^{-1}) = -1x^{-2}$, which is $\frac{-1}{x^2}$. It works!

What is the derivative of $5x^4$? Bring the 4 down and reduce by 1 to get $4 \cdot 5x^3 = 20x^3$. Finding this derivative using limits would have taken a very long time.

Derivative of a Polynomial

Once you know the pattern to differentiate a single power of x, polynomials that are sums and differences of monomials are no problem. The derivative of a sum or difference is the sum or difference of the derivatives of the terms taken independently. This is because the limit of a sum is the sum of the limits taken independently, and a derivative is just a limit. In more formal terms, if $y = f(x) + g(x)$, then $\frac{dy}{dx} = f'(x) + g'(x)$. One quick example should be enough to illustrate this point.

EXAMPLE 5-5

Find the derivative of $y = 2x^4 - 6x^3 + 11x^2 - 5x + 9$.

The new pattern for differentiating is emphasized in the following line of work.

$$\frac{dy}{dx} = 4 \cdot 2x^{4-1} - 3 \cdot 6x^{3-1} + 2 \cdot 11x^{2-1} - 1 \cdot 5x^{1-1} + 0$$

$$\frac{dy}{dx} = 8x^3 - 18x^2 + 22x - 5$$

Derivatives of Products and Quotients

Sums and differences can be differentiated easily, but unfortunately this is not the case with products and quotients. First take a look at products. A simple counterexample should confirm that something else is happening with products. For a moment, assume that the derivative of a product is the product of each derivative. Under this assumption, the derivative of $f(x)=x^3 \cdot 5x^4$ would be $f'(x)=3x^2 \cdot 20x^3$. The properties of exponents simplifies this to $f'(x)=60x^5$. Now do the problem using the properties of exponents first! The original function $f(x)=x^3 \cdot 5x^4=5x^7$. In this case, $f'(x)=35x^6$. Clearly, then, the derivative of a product is *not* the product of each derivative.

The actual rule is a bit more complicated, and proving the rule for products (and quotients) is beyond the scope of this book. If you are interested, you can look it up online. Just search for "proof of the product rule."

The Product Rule: Let $f(x)=p(x) \cdot q(x)$, where p and q are differentiable functions. Then $f'(x)=p'(x) \cdot q(x)+q'(x) \cdot p(x)$. In words, one might say, "The derivative of the first function times the second function plus the derivative of the second function times the first function." Essentially, functions are "polite," and each function "waits its turn" while the other function gets differentiated.

You may be wondering why you wouldn't just multiply out all the functions first and then differentiate. But remember that you will not always be working with products of functions that can be simplified, so this rule is necessary.

If you keep the "politeness concept" in mind, the product rule can be extended to more than two multiplied functions. All other functions are grouped with the derivative of one function. For example, if a, b, and c are all functions, and $y=a \cdot b \cdot c$, then $y'=a' \cdot b \cdot c+b' \cdot a \cdot c+c' \cdot a \cdot b$. Example 5-6 and Example 5-7 find the derivative of a product by expanding and then differentiating and by using the product rule, respectively. The results should be the same!

EXAMPLE 5-6

Find the derivative of $y = x^3(4x^2 + 3x)$ by distributing and then differentiating the result.

$$y = 4x^5 + 3x^4$$

$$\frac{dy}{dx} = 20x^4 + 12x^3$$

EXAMPLE 5-7

Find the derivative of $y = x^3(4x^2 + 3x)$ by using the product rule and then simplifying.

$$\frac{dy}{dx} = 3x^2(4x^2 + 3x) + (8x + 3) \cdot x^3$$

$$\frac{dy}{dx} = 12x^4 + 9x^3 + 8x^4 + 3x^3$$

$$\frac{dy}{dx} = 20x^4 + 12x^3$$

Note that in this case, using the product rule first turned out to be more work, but that is not always true. As you gain more experience, deciding whether to use one method or another will become easier.

The Quotient Rule

Of course, if there is a special rule for products, there is also a special rule for quotients of functions.

The Quotient Rule: Let $f(x)=\frac{p(x)}{q(x)}$, where p and q are differentiable functions. Then $f'(x)=\frac{q(x)\cdot p'(x)-p(x)\cdot q'(x)}{[q(x)]^2}$. A mnemonic device can be useful here. If you think of "low" as the denominator, of "high" as the numerator, of "d(low)" as "the derivative of the denominator," and of "d(high)" as "the derivative of the numerator," then a quick phrase to spur your recall is "low d(high) minus high d(low) all over low squared."

However you choose to remember the rule, there is no substitute for plenty of practice applying it. Example 5-8 is just a first step in mastering this derivative pattern.

EXAMPLE 5-8

Find the derivative of $y=\frac{x^2}{7x+9}$.

$$\frac{dy}{dx}=\frac{(7x+9)\cdot\frac{d}{dx}(x^2)-x^2\cdot\frac{d}{dx}(7x+9)}{(7x+9)^2}$$

Compare the pattern to the rule in the sidebar.

$$\frac{dy}{dx}=\frac{(7x+9)\cdot(2x)-x^2\cdot 7}{(7x+9)^2}$$

Differentiate x^2 and $(7x+9)$.

$$\frac{dy}{dx}=\frac{14x^2+18x-7x^2}{(7x+9)^2}$$

Distribute $2x$.

$$\frac{dy}{dx}=\frac{7x^2+18x}{(7x+9)^2}$$

Combine like terms.

Many times, the square of the denominator does not need to be expanded. You may sometimes need to expand the denominator if you are trying to match your answer to some multiple-choice options.

As you move through calculus and accumulate more and more knowledge, try to guard against making the common mistake of forgetting to use rules you learned early, such as the product rule and the quotient rule. Any time you are finding a derivative, ask yourself whether the product rule or the quotient rule applies. Never differentiate individual factors of the variable or individual numerators and denominators.

Skill Check

If you do not feel ready to move into practice, review the chapter again, focusing on the patterns for the various differentiation rules. All of the problems in this Skill Check can be done without a calculator. Check your answers against the solutions in Appendix D. Finding additional practice online is a little harder for this topic, but not impossible. You may encounter derivatives of functions not covered up to this point. You may also find that worked-out examples are abundant, but problems to try by yourself and then check are more scarce.

1. Use the "x approaches a" definition of the derivative to find the slope of $y=\frac{1}{2}x^2$ at $x=4$. Check your answer by using the derivative rule for powers of x.
2. $\lim\limits_{x\to 7}\frac{\left(2/x\right)-\left(2/7\right)}{x-7}$ finds the slope of what function, at what point?
3. Write the equation of the line tangent to $y=x^3-3x+5$ at $x=1$.
4. Given $h(x)=(x^2+9x-11)(5x^3-7x^2+13)$. Write the first step in using the product rule to find $h'(x)$. Do not do all the algebra to simplify beyond the first step.
5. Given $g(x)=\frac{12x^3-8x}{2x}$. Find $\frac{dg}{dx}$ by first reducing the fraction and then differentiating.
6. Given $g(x)=\frac{12x^3-8x}{2x}$. Find $\frac{dg}{dx}$ by using the quotient rule. Then show that your answer is equivalent to the result in Problem 5.

CHAPTER 6

Derivatives of Trigonometric Functions

Now it's time to expand your calculus world beyond polynomials and rational functions. There are a multitude of applications of trigonometric functions, such as hours of daylight through the year, sound waves, and alternating currents, to name just a few. Measuring rates of change in these functions is important to anyone who works in meteorology, audiology, electronics, or any of a number of other fields. This chapter will introduce the derivatives of the six basic trigonometric functions, which are so useful in real-life applications.

Derivatives on a Graphing Calculator

Since the early 1990s, graphing calculators have had a profound impact on how calculus has been taught and learned. Before the advent of graphing calculators, most students learned calculus as a series of symbolic manipulations and had limited ability to visualize what they were doing. A fortunate minority had access to computer technology that afforded them some visual reinforcement of abstract concepts, such as local linearity. But those days are long gone, and used properly, a graphing calculator can be a powerful tool for learning. Currently, Texas Instruments and Casio calculators dominate the market. The references in this text will be to the syntax of a TI-84 model calculator. If you have this type of calculator, you should try all the examples given here. Even though a TI-84 is not a computer algebra system, and so does not work symbolically, it can do powerful numerical calculus. For now, you will learn three calculus skills with the calculator. It will be assumed that you have a basic familiarity with the calculator.

In calculus, all calculators work with trigonometric functions done in radian mode. Whatever brand of calculator you own, if you are not already in radian mode, set it to radian mode now. Calculating derivatives of trigonometric functions in degree mode will result in wrong answers.

Derivative on the Home Screen

The screen you see when you turn on the calculator is called the Home screen. On this screen you can find a very accurate, but not always perfect, value of the slope of a function at a given point. Press the button MATH, followed by 8 to get a command called nDeriv(. This can also be found in the catalogue. The syntax that follows the "open parentheses" symbol is: nDeriv(function, independent variable, value where you want the derivative). Thus, to find the derivative of $y = x^2$ evaluated at $x = 3$, you type nDeriv(x^2, x, 3). Because you know that the derivative of x^2 is $2x$, you should get 6, as shown in Figure 6-1. What the calculator does to get this value is calculate the slope of the secant between the points one one-thousandth on either side of your desired number. This is why there is sometimes a tiny amount of error.

Figure 6-1

Derivative on the Graph Screen

You can also find a slope of a function from its graph. Graph the function $y=\sqrt{x}$ in a window with $-2 \leq x \leq 10$. Now press the 2nd button followed by TRACE. This accesses the menu called CALC. The sixth choice is $\frac{dy}{dx}$. Choose it by arrowing down and pressing ENTER or by pressing the number 6. Now type a 4 and press ENTER. The result on the screen shows the value of the derivative of $y=\sqrt{x}$ at $x=4$. One limitation of this method is you must get a good viewing window for your graph, and you can type in x values only in your graphing window. Figure 6-2 shows a series of the screen shots for this procedure.

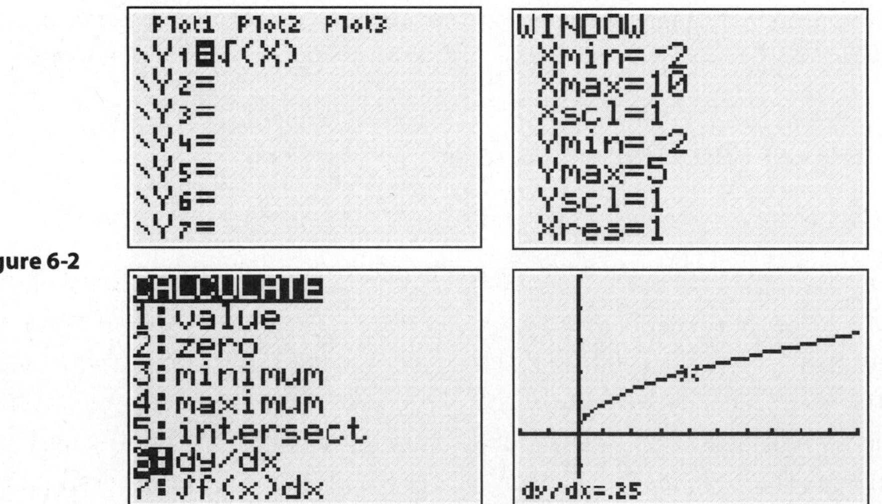

Figure 6-2

Graphing a Derivative as a Function

Perhaps one of the most powerful capabilities of the calculator is to actually graph the derivative as a function by calculating many numerical derivatives and plotting them. Go to the place where you normally graph functions. Position the cursor in the Y1 location, clear out any remaining functions, and get nDeriv(from the MATH menu or the catalogue. Finish typing so it looks like Y1 = nDeriv($x^2 - 3x, x, x$). Press ZOOM followed by 6 to graph in a standard –10 to 10 window. Do you see a line? From the function whose derivative you plotted, do you know the equation of the line? It should be $y = 2x - 3$ because the derivative of $x^2 - 3x$ is $2x - 3$. Figure 6-3 shows a series of screen shots from this procedure.

Figure 6-3

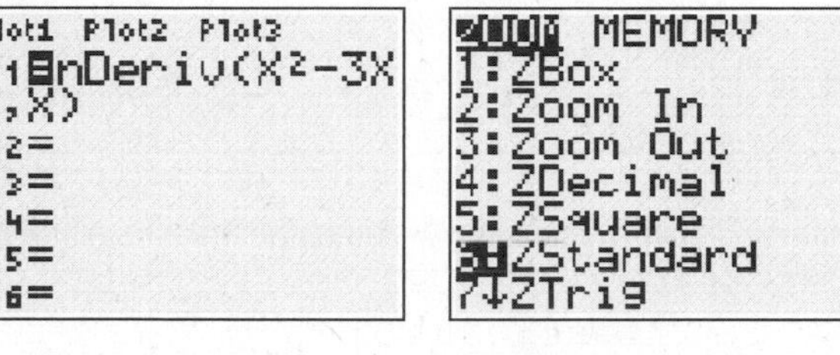

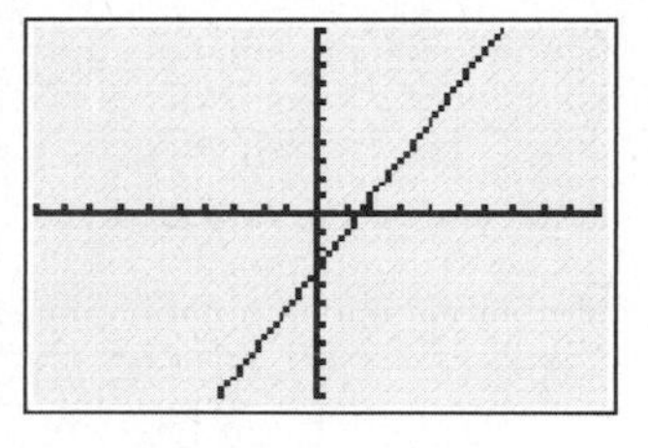

There is much more that can be done on a graphing calculator to help you learn calculus, but this text is not intended to be a calculator manual. As skills are needed, they will be taught so that the calculator will act as an appropriate supplement to the calculus you are learning.

Derivative of the Sine Function

Because the sine function is your entry into the realm of derivatives of trigonometric functions, its derivative will be justified in a couple of ways. One way to justify the derivative of the sine curve is to use your graphing calculator to find the value of the derivative at a variety of x values, record them, and then plot each of those x values paired with its corresponding derivative value as the y value. In this way you are building a set of ordered pairs of the form $[x, f'(x)]$. But for the sake of efficiency, you could use the method of graphing a derivative that was just presented. In Y1 enter: nDeriv(sin(x), x, x). Press ZOOM followed by 6. Again, this window is –10 to 10 in each direction. Be patient. The calculator is doing ninety-four derivatives for you! If you know how, manually make the y values smaller to get a better view of the graph. In Figure 6-4, the y values are set from –2 to 2.

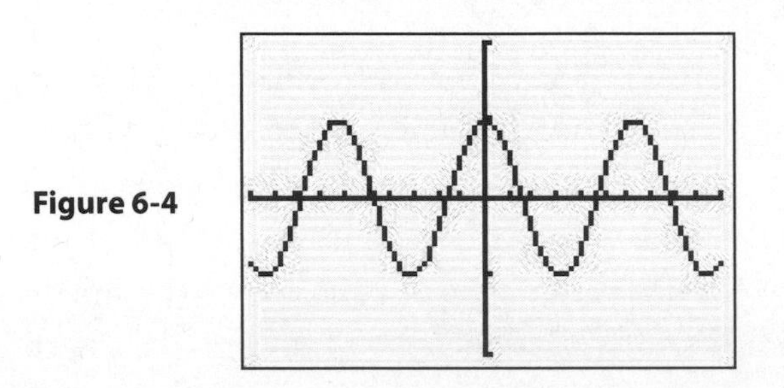

Figure 6-4

You are looking at a graph of the derivative of $\sin(x)$. A good trigonometry student will recognize the graph. Do you? It is, of course, the cosine function!

A somewhat stronger justification for this can be done with the definition of the derivative. Example 6-1 takes you through the steps. Two of the limits you encounter will be determined by graphing on your calculator.

EXAMPLE 6-1

Find the derivative of the function $f(x)=\sin(x)$.

$$f'(x)=\lim_{h\to 0}\frac{f(x+h)-f(x)}{h}$$

$$f'(x)=\lim_{h\to 0}\frac{\sin(x+h)-\sin(x)}{h}$$

$$f'(x)=\lim_{h\to 0}\frac{\sin(x)\cos(h)+\cos(x)\sin(h)-\sin(x)}{h}$$

$$f'(x)=\lim_{h\to 0}\frac{\sin(x)\cos(h)-\sin(x)}{h}+\frac{\cos(x)\sin(h)}{h}$$

$$f'(x)=\lim_{h\to 0}\sin(x)\cdot\frac{\cos(h)-1}{h}+\lim_{h\to 0}\cos(x)\cdot\frac{\sin(h)}{h}$$

By graphing, $\lim_{h\to 0}\frac{\cos(h)-1}{h}=0$ and $\lim_{h\to 0}\frac{\sin(h)}{h}=1$

$f'(x)=\lim_{h\to 0}[\sin(x)\cdot 0]+\lim_{h\to 0}[\cos(x)\cdot 1]$

The first limit goes to zero, and the second limit becomes $\cos(x)\cdot 1=\cos(x)$. The conclusion is that if $y=\sin(x)$, then $\frac{dy}{dx}=\cos(x)$.

Derivative of the Cosine Function

You might make an educated guess that if the derivative of the sine function is the cosine function, then the derivative of the cosine function should be the sine function. That is a logical thought, but it is not quite correct. Before we give it away too quickly, you can strengthen your graphical thinking skills by following the discussion of Figure 6-5. Thinking graphically is actually a critical skill that will get a lot more work in a later chapter, so developing a foundation for it here will serve you well as you progress.

Figure 6-5

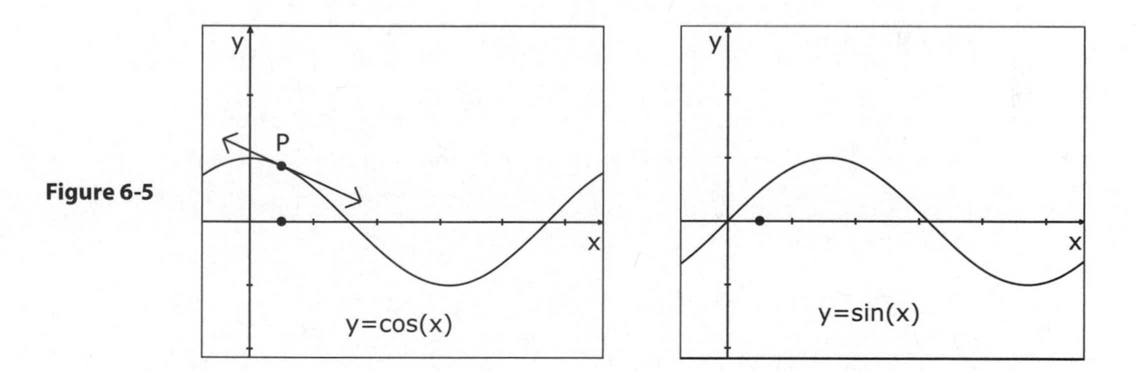

The graph on the left is $y=\cos(x)$ with a tangent line drawn at point P. The graph on the right is $y=\sin(x)$. The x value of point P is marked on each axis. Would the value of the derivative of the cosine at point P be positive or negative? Because the tangent line has a downward slope to the right, the derivative would be negative. If the sine function were the derivative of the cosine function, then at that same x value the sine should be negative, but it is not!

So now it is time to use your graphing calculator to draw your own conclusion. Graph the derivative of the cosine function with Y1 = nDeriv(cos(x), x, x). Do you recognize the graph? If not, graphing Y2 = sin(x) should give you a clue. The derivative of the cosine function is the opposite of the sine function. The sidebar summarizes the first two conclusions in succinct symbol form.

Derivatives of Sine and Cosine: For all real numbers x, if $f(x) = \sin(x)$, then $f'(x) = \cos(x)$.

For all real numbers x, if $g(x) = \cos(x)$, then $g'(x) = -\sin(x)$.

Derivative of the Tangent Function

Now that you know the derivatives of sine and cosine, you can develop the derivative of the tangent function by using the quotient rule. Note, though, that there will be places in the domain where the derivative of the tangent function is undefined. At every odd multiple of $\frac{\pi}{2}$, the tangent graph has vertical tangents because its denominator, cosine, is zero. Example 6-2 shows the derivation of the derivative of the tangent function.

EXAMPLE 6-2

If $f(x) = \tan(x)$, find $f'(x)$.

$$\tan(x) = \frac{\sin(x)}{\cos(x)}$$

$$\frac{d}{dx}[\tan(x)] = \frac{d}{dx}\left[\frac{\sin(x)}{\cos(x)}\right]$$

$$\frac{d}{dx}[\tan(x)] = \frac{\cos(x)\frac{d}{dx}[\sin(x)] - \sin(x)\frac{d}{dx}[\cos(x)]}{[\cos(x)]^2}$$

$$\frac{d}{dx}[\tan(x)] = \frac{\cos(x)\cdot\cos(x) - \sin(x)\cdot[-\sin(x)]}{\cos^2(x)}$$

$$\frac{d}{dx}[\tan(x)] = \frac{\cos^2(x) + \sin^2(x)}{\cos^2(x)}$$

$$\frac{d}{dx}[\tan(x)] = \frac{1}{\cos^2(x)} = \sec^2(x)$$

It is technically correct to use $\frac{1}{\cos^2(x)}$ as the derivative of the tangent, but $\sec^2(x)$ is the much more commonly accepted form.

With three derivatives of the six basic trigonometric functions established, take a moment to try using some of this new knowledge in the next example. For better practice, see if you can do the problem yourself before looking at the solution.

EXAMPLE 6-3

Find the slope of the line that is tangent to the graph of $y = x^2\cos(x)$ at $x = \frac{\pi}{2}$.

$$\frac{dy}{dx} = \frac{d}{dx}(x^2)\cdot\cos(x) + \frac{d}{dx}[\cos(x)]\cdot x^2$$ Remember the product rule!

$$\frac{dy}{dx} = 2x\cdot\cos(x) + [-\sin(x)]\cdot x^2$$

$$\left.\frac{dy}{dx}\right|_{x=\pi/2} = 2\left(\frac{\pi}{2}\right)\cdot\cos\left(\frac{\pi}{2}\right) - \sin\left(\frac{\pi}{2}\right)\cdot\left(\frac{\pi}{2}\right)^2$$ Evaluate at $x = \frac{\pi}{2}$.

$$\left.\frac{dy}{dx}\right|_{x=\pi/2} = \pi\cdot 0 - 1\cdot\left(\frac{\pi}{2}\right)^2 = -\frac{\pi^2}{4}$$

Derivatives of Other Trigonometric Functions

As you know, there are three more basic trigonometric functions besides cosine, sine, and tangent. Their reciprocal functions are secant, cosecant, and cotangent, respectively. The derivatives are given here without proof.

- $\frac{d}{dx}[\sec(x)] = \sec(x) \cdot \tan(x)$
- $\frac{d}{dx}[\csc(x)] = -\csc(x) \cdot \cot(x)$
- $\frac{d}{dx}[\cot(x)] = -\csc^2(x)$

Six trigonometric derivatives may seem like a lot to remember, but there are a few small patterns that may help. As you examine the derivatives, note that the derivative of every function with the prefix "co" has a negative derivative, whereas the others are positive. Also, the derivatives of cofunctions involve the cofunctions of the derivatives. That may sound like double-talk, but (for example) the derivative of tangent is secant squared, and the derivative of cotangent involves cosecant, the cofunction of the tangent's derivative.

In electric circuits, apparent power is important for determining heat, which affects wire size. When an electric circuit has power *P* and an impedance phase angle *Q*, apparent power is defined by $A = P\sec(Q)$. If electrical engineers want to know the rate of change of the apparent power with respect to the phase angle, they work with the derivative of the apparent power equation.

These derivative rules are definitely facts you need to have at your disposal, so be sure to commit them to memory. If it helps you, make a set of flash cards with the function being differentiated on the front and the resulting derivative on the back. It's that important!

Using Trigonometric Identities

There is one more subtlety to be noted when working with trigonometric functions. Before you differentiate a function, especially when it looks complicated, consider using trigonometric identities to simplify the expression. The derivatives of both expressions will be equivalent, but arriving at the answer is significantly less difficult when you work with a simplified form. Of course, sometimes there is no easy way to simplify a trigonometric expression, and differentiating it will just be a lot of work. Compare Examples 6-4 and 6-5, which show the same problem done both ways.

EXAMPLE 6-4

Find the derivative of $g(x) = \dfrac{\sin(x)}{\tan(x)}$.

$$g'(x) = \frac{\tan(x) \cdot \frac{d}{dx}[\sin(x)] - \sin(x) \cdot \frac{d}{dx}[\tan(x)]}{[\tan(x)]^2}$$

Use the quotient rule.

$$g'(x) = \frac{\tan(x) \cdot \cos(x) - \sin(x) \cdot \sec^2(x)}{\tan^2(x)}$$

$$g'(x) = \frac{\frac{\sin(x)}{\cos(x)} \cdot \cos(x) - \sin(x) \cdot \sec^2(x)}{\tan^2(x)}$$

Simplify as much as possible.

$$g'(x) = \frac{\sin(x) - \sin(x) \cdot \sec^2(x)}{\tan^2(x)}$$

$$g'(x) = \frac{\sin(x)[1 - \sec^2(x)]}{\tan^2(x)}$$

$$g'(x) = \frac{\sin(x) \cdot [-\tan^2(x)]}{\tan^2(x)} = -\sin(x)$$

EXAMPLE 6-5

Find the derivative of $g(x)=\dfrac{\sin(x)}{\tan(x)}$, but simplify the expression first.

$$g(x)=\sin(x)\cdot\frac{1}{\tan(x)}$$

$$g(x)=\sin(x)\cdot\frac{\cos(x)}{\sin(x)}$$

$$g(x)=\cos(x)$$

$$g'(x)=-\sin(x)$$

Even with a few additional steps showing details involved in simplifying the original problem, Example 6-5 shows how much easier this is to simplify before differentiating. That may not always be the case, but you should at least check, because you might save yourself a lot of work. The usefulness of simplifying with trigonometric identities reinforces the importance of knowing the primary identities given in Appendix A.

Skill Check

Try to have the derivatives of the six basic trigonometric functions memorized before you work through this Skill Check. It will also be useful to be able to recall unit circle values of trigonometric functions. You should not use a calculator unless the problem instructs you to do so.

1. Given $y=\sin(x)\cdot\cos(x)$. Work on the Home screen of your calculator to find the value of $y'(2)$.
2. Without taking the derivative by hand, use your calculator to produce a graph of the derivative of $f(x)=\dfrac{x}{x^2+1}$.
3. Find the equation of the line that is tangent to $y=\cos(x)$ at $x=\dfrac{\pi}{2}$.
4. Given $h(x)=\cos(x)\cdot\csc(x)$. Find $h'(x)$.

5. Use the quotient rule to show that the derivative of $\sec(x)$ is $\sec(x)\cdot\tan(x)$.

6. The Figure 6-6 shows the graph of $y=2+\cos(x)$ on $[0,2\pi]$. On the graph, place an X where the graph appears to have the largest positive slope. Then use calculus to build an argument that this occurs at $x=\frac{3\pi}{2}$.

Figure 6-6

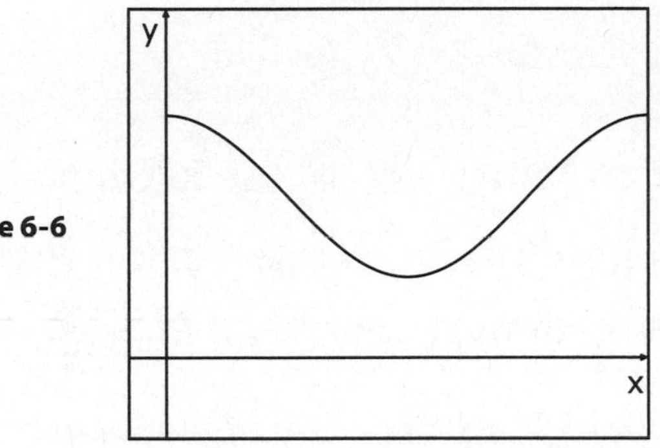

CHAPTER 7

The Chain Rule

The concept presented in this chapter is really the fulcrum of any introductory calculus course. The chain rule is necessary in order to deal with significantly more advanced functions. Without it, your journey through calculus would be like trying to take a 100-mile trip on a tricycle. The chain rule is also used to develop many other major ideas. The chain rule is a critical concept for the rest of your study of calculus.

Derivatives of Composite Functions

Perhaps you have noticed that so far, every function you have worked with has been a simple function of an independent variable, not a composition of several functions. For example, you have worked with $y=\cos(x)$, but not with $y=\cos(5x)$. The reason for this is simply that finding a derivative of a composite function is a bit more involved. A natural, but incorrect, guess at the derivative of $y=\cos(5x)$ is $\frac{dy}{dx}=-\sin(5x)$.

Try to discover the correct derivative by using the calculator as a tool for learning. Use the technique from the last chapter for graphing a derivative. Let $\text{Y1}=\text{nDeriv}(\cos(5x),x,x)$, and examine the graph in a standard –10 to 10 window by pressing ZOOM 6. The graph looks like $y=-\sin(x)$, but it has a different amplitude. The amplitude is actually 5, which means that the derivative of $y=\cos(5x)$ is $\frac{dy}{dx}=-5\sin(5x)$. The real question is "Where did that extra 5 come from?" Understanding the answer to the question is intertwined with understanding the different types of notation used by famous mathematicians such as Gottfried Wilhelm Leibniz and Joseph-Louis LaGrange.

Any variable can be differentiated "with respect to" another variable. It just means you are examining the ratio of an infinitely small change in one variable to an infinitely small change in another variable. Being comfortable with this idea will be helpful as you encounter more applications. For instance, $\frac{dV}{dt}$ may stand for the rate of change of volume with respect to time.

Understanding the Notation

Leibniz introduced the notation of infinitesimals. He thought of the small changes in x and y geometrically and concluded that under the right conditions, $\frac{dy}{dx}=\frac{dy}{du}\cdot\frac{du}{dx}$ even though they cannot actually be treated the same way as variables in algebra. Nonetheless, applying this notation to finding the derivative of a composite function can be very informative. The function $y = \cos(5x)$ is the composition of two noncomposite functions, $y = \cos(u)$ and $u = 5x$. Each one can be differentiated with respect to its own independent variable to give $\frac{dy}{du} = -\sin(u)$ and $\frac{du}{dx} = 5$. Using Leibniz's notation, $\frac{dy}{dx} = \frac{dy}{du} \cdot \frac{du}{dx}$ means $\frac{dy}{dx} = -\sin(u) \cdot 5$. Using a simple substitution from $u = 5x$, the derivative becomes $\frac{dy}{dx} = -5\sin(5x)$.

Joseph-Louis LaGrange used the "prime" notation that was introduced in previous chapters. It is a little less intuitive, but it provides another perspective on this important rule and actually leads to a useful informal approach to the chain rule. If y is a composite function of the form $y = f(g(x))$, then $y' = f'(g(x)) \cdot g'(x)$. Think of a composite function as having an outer function and an inner function. To differentiate a composite function, take the derivative of the outer function, without changing the inner function, and multiply it by the derivative of the inner function. For $y = \cos(5x)$, the outer function is the cosine function, and the inner function is $5x$. Therefore, differentiate the cosine without changing the $5x$, and then multiply it by the derivative of $5x$, which results in $y' = -\sin(5x) \cdot 5$. It is important to be familiar with both the Leibniz and the LaGrange notations, because they are often used interchangeably in calculus readings.

Chain Rule for Composite Functions: Let $y = f(h(x))$ be a composite function of h and f. If h is differentiable at x and f is differentiable at $h(x)$, then the derivative of y is $y' = f'(h(x)) \cdot h'(x)$. Using Leibniz's notation, if $y = f(u)$ and $u = h(x)$, then y is a function of u and it makes sense to find $\frac{dy}{du}$. Likewise, u is a function of x, and it makes sense to find $\frac{du}{dx}$. Therefore, $\frac{dy}{dx} = \frac{dy}{du} \cdot \frac{du}{dx}$.

Thinking of a composite function by identifying outer and inner functions is often referred to as using a layering rule. It is a slightly less formal way to approach differentiating, but it can be a time saver as your ability to find derivatives increases. Examples 7-1 and 7-2 illustrate the application of the two methods to the same differentiation problem.

EXAMPLE 7-1

Find the derivative of $y = \tan(x^3 + 1)$ using Leibniz's notation.

Let $y = \tan(u)$ and $u = x^3 + 1$.

Then $\frac{dy}{du} = \sec^2(u)$ and $\frac{du}{dx} = 3x^2$.

Thus $\frac{dy}{dx} = \frac{dy}{du} \cdot \frac{du}{dx} = \sec^2(u) \cdot 3x^2$.

$$\frac{dy}{dx} = 3x^2 \cdot \sec^2(x^3 + 1)$$

EXAMPLE 7-2

Find the derivative of $y = \tan(x^3 + 1)$ using LaGrange's notation.

The outer function, tangent, is differentiated without changing the inner function, $x^3 + 1$. The result is then multiplied by the derivative of the inner function.

$$y' = \sec^2(x^3 + 1) \cdot \frac{d}{dx}(x^3 + 1)$$

$$y' = \sec^2(x^3 + 1) \cdot 3x^2$$

The challenge with the layering approach is that sometimes layers can get deeper than two functions, and for students new to calculus, it is easy to lose track of a step. Leibniz's approach involves writing down each function and keeps track a bit better.

Applying the Chain Rule to Powers

You just learned how the chain rule can be used to find the derivative of a trigonometric function when the angle has a derivative of something other than 1. Another situation that requires the chain rule arises when you encounter powers of quantities such as binomials, trinomials, trigonometric functions, or even logarithmic functions. As previously mentioned, skill with decomposing functions is very useful. In this situation, seeing the exponent as the outer function and the base as the inner function is the key. To convince you that the process works as it should, Examples 7-3 and 7-4 differentiate the same power in two different ways.

EXAMPLE 7-3

Find the derivative of $y=(3x+1)^2$ by first expanding the binomial square and then differentiating the resulting polynomial.

$y=(3x+1)(3x+1)$

$y=9x^2+6x+1$ Expand by the "FOIL" method.

$\frac{dy}{dx}=18x+6$ Differentiate each power of x.

EXAMPLE 7-4

Find the derivative of $y=(3x+1)^2$ using the chain rule.

$y=u^2$ and $u=3x+1$ Set up to use Leibniz's notation.

$\frac{dy}{du}=2u$ and $\frac{du}{dx}=3$ Differentiate each equation.

$\frac{dy}{dx}=\frac{dy}{du}\cdot\frac{du}{dx}=2u\cdot 3$ Multiply to find $\frac{dy}{dx}$.

$\frac{dy}{dx}=6u=6(3x+1)$ Substitute for u.

Note that the two derivatives are equivalent. They would not have been equal without the additional factor of 3 obtained from the chain rule. Example 7-3 also lends itself to being done either way, because it is relatively easy to expand the power to its polynomial form. Would you want to expand the power if the task were to differentiate $y=(x^2+5x-9)^{11}$?

As you become more accomplished at using the chain rule, thinking in terms of the layering principle can save some time, but be sure that you do not use it at the expense of accuracy! To use it on a power, think about applying the power rule to the quantity, without changing the base. That is differentiating the "outer layer." Then multiply the result by the derivative of the base, the "inner layer." The next example takes you through this process, but don't be too hasty to use it until you are getting most of your derivatives correct using a more organized approach.

EXAMPLE 7-5

Find the derivative of $y=(x^2+5x-9)^{11}$.

The outer layer is the power of 11. The inner layer is the trinomial.

Apply the power rule without changing the base, and prepare to differentiate the inner layer.

$$\frac{dy}{dx}=11(x^2+5x-9)^{11-1}\frac{d}{dx}(x^2+5x-9)$$

$$\frac{dy}{dx}=11(x^2+5x-9)^{10}(2x+5)$$

Many times in calculus, answers may be left without being simplified, because the greater interest is in doing the calculus correctly, not spending a lot of time on algebraic manipulation. In multiple-choice testing situations, however, simplifying a result may be necessary if the correct answer choice is not in the same form as your answer.

A Longer Chain!

With more complicated functions, it may be necessary to decompose them into three or more components. In other words, the composite function is more than just two layers. For instance, $y = \sqrt{\sin(x^3 + 1)}$ has three layers. You have a square root function as the outermost layer. There is a sine function inside the square root function. Finally, a cubic function lurks inside the sine function. Patience and organization are the key to success. It is also important to remember that the final derivative should be presented in terms of the original independent variable, x.

EXAMPLE 7-6

Find the derivative of $y = \sqrt{\sin(x^3 + 1)}$.

$y = \sqrt{u}$ and $u = \sin(p)$ and $p = x^3 + 1$ — Decompose the function into three components.

$$\frac{dy}{du} = \frac{1}{2}u^{\left(\frac{-1}{2}\right)}$$

Remember that $\sqrt{u} = u^{\left(\frac{1}{2}\right)}$.

Bring the $\frac{1}{2}$ down and reduce by 1.

$$\frac{du}{dp} = \cos(p) \text{ and } \frac{dp}{dx} = 3x^2$$

$$\frac{dy}{dx} = \frac{dy}{du} \cdot \frac{du}{dp} \cdot \frac{dp}{dx}$$

$$\frac{dy}{dx} = \frac{1}{2}u^{\left(\frac{-1}{2}\right)} \cdot \cos(p) \cdot 3x^2$$

Substitute for the derivatives.

$$\frac{dy}{dx} = \frac{1}{2}[\sin(p)]^{\left(\frac{-1}{2}\right)} \cdot \cos(x^3 + 1) \cdot 3x^2$$

Substitute for p in terms of x and for u in terms of p.

$$\frac{dy}{dx} = \frac{1}{2\sqrt{\sin(x^3+1)}} \cdot \cos(x^3+1) \cdot 3x^2$$

Substitute for p in terms of x, and write the $\frac{-1}{2}$ power as a square root in the denominator.

With continual advances in computer technology, implicit functions are playing a key role in many industries. Implicit functions programmed into advanced drawing and design software can be used to produce three-dimensional imaging in everything from the manufacturing industry to animated movies.

Implicit Differentiation

Not all functions can be easily written in a form where the dependent variable is completely isolated from the dependent variable, and for some relations it is simply not possible. Functions and relations of these types are called implicit. Most standard graphing calculators cannot even produce a graph of an implicit relation, but fortunately, many computer programs exist to help with the task.

Naturally, whether an implicit relation is a function or not, if it can be plotted on the Cartesian coordinate plane, then its slope (or rate of change) can be examined. By this point in your studies, when you see the word *slope*, you should think instantly of a derivative. The key to finding a derivative of an implicit relation is to remember that the y values are interdependent with the x values, so whether or not it can be isolated, y is a function of x. This means that every time a term with a y in it is differentiated, the chain rule requires a $\frac{dy}{dx}$. The other subtlety of y being a function of x is that any product involving both x and y in the implicit equation requires using the product rule when differentiating. The first example that follows uses a function that can be examined implicitly and explicitly, thereby providing an opportunity to confirm the process of implicit differentiation. The graph of the relation appears in Figure 7-1.

Figure 7-1

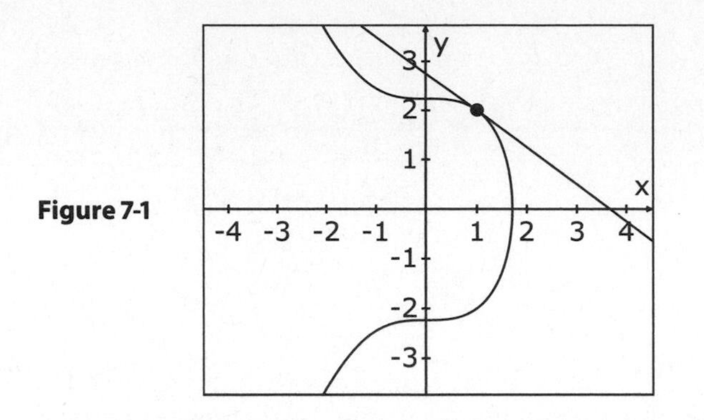

EXAMPLE 7-7

Find the equation of the line tangent to $x^3 + y^2 = 5$ at the point (1,2).

$\frac{d}{dx}(x^3) + \frac{d}{dx}(y^2) = \frac{d}{dx}(5)$ Differentiate each term with respect to x.

$3x^2 + 2y\frac{dy}{dx} = 0$ The power rule was used on y^3, and $\frac{dy}{dx}$ comes from the derivative of the base, y, with respect to x.

$3(1)^2 + 2(2)\frac{dy}{dx} = 0$ Substitute (1,2).

$$\frac{dy}{dx} = \frac{-3}{4}$$

$y - 2 = \frac{-3}{4}(x - 1)$ The tangent line is shown in Figure 7-1.

EXAMPLE 7-8

Find the equation of the line tangent to $x^3 + y^2 = 5$ at the point (1,2) by first writing y as a function of x.

$y = \pm\sqrt{5 - x^3}$ — Subtract x and square-root both sides of the equation. The positive square root contains (1,2) and will be used for differentiating.

$y = (5 - x^3)^{(1/2)}$

$\frac{dy}{dx} = \frac{1}{2}(5 - x^3)^{(-1/2)} \cdot (-3x^2)$ — Power rule was used and $-3x^2$ is the derivative of the base.

$\frac{dy}{dx} = \frac{1}{2}(5 - 1^3)^{(-1/2)} \cdot (-3(1)^2) = \frac{-3}{4}$ — Substitute $x = 1$.

$y - 2 = \frac{-3}{4}(x - 2)$

As you can see, the same result was obtained by both methods. But note that the implicit differentiation method produced a derivative expression that involved both x and y, which meant that the entire ordered pair was used to produce the slope value. If y can be written as an explicit function of x, then the derivative expression depends solely on x. Of course, there are times when neither x nor y can be isolated, and at those times, implicit differentiation is necessary.

EXAMPLE 7-9

Find the derivative of $x + xy^3 = yx^3$.

$\frac{d}{dx}(x) + \frac{d}{dx}(xy^3) = \frac{d}{dx}(yx^3)$ — Prepare to differentiate each term of the equation.

$1 + x \cdot 3y^2 \frac{dy}{dx} + y^3 \cdot 1 = y \cdot 3x^2 + x^3 \cdot \frac{dy}{dx}$ — Product rule was used on xy^3 and xy^3. The $\frac{dy}{dx}$ showed up each time a y term was differentiated.

$3xy^2\frac{dy}{dx} - x^3\frac{dy}{dx} = 3x^2y - y^3 - 1$ Isolate all $\frac{dy}{dx}$ terms on one side of the equation.

$(3xy^2 - x^3)\frac{dy}{dx} = 3x^2y - y^3 - 1$ Factor out $\frac{dy}{dx}$ as a greatest common factor.

$\frac{dy}{dx} = \frac{3x^2y - y^3 - 1}{3xy^2 - x^3}$ Divide both sides of the equation by $3xy^2 - x^3$.

Skill Check

It is time once again to test your skills. If necessary, review the derivatives of the various functions, as well as the product rule, quotient rule, and chain rule, before you begin. As you can see, calculus constantly builds on what has already been learned. Leaving earlier knowledge behind as you move forward is a bad idea!

1. Find $\frac{dy}{dx}$ if $y = \sqrt[3]{x^3 + 6x}$.

2. Given $p = \tan(w)$ and $w = \cos(t)$, use Leibniz notation to find $\frac{dp}{dt}$. The final answer should be all in terms of t.

The following table shows functions f and g and their derivative values at x = 1 and x = 2. Use the information in the table to answer Problems 3 and 4.

x	$f(x)$	$g(x)$	$f'(x)$	$g'(x)$
1	3	2	0	$\frac{3}{4}$
2	5	-4	$\frac{1}{3}$	-1

3. If $y = \frac{f(x)}{g(x)}$, find the value of $y'(2)$.

4. If $y = f(g(x))$, find the value of $y'(1)$.

5. If $y^3 = x^2 + \sin(y)$, find $\frac{dy}{dx}$.

6. Find the equation of the line tangent to the graph of $x^2y + y^3 + x^3 = 13$ at the point $(1,2)$.

CHAPTER 8

Derivatives of Other Functions

Because many natural phenomena change according to patterns other than polynomials and trigonometric functions, it will be useful for you to know how to find derivatives of some of these other functions. Two common functions that occur quite often in nature are exponential and logarithmic functions. Knowing how to take their derivatives will enable you to examine the rates of change in bacteria colony populations, tissue growth rates in the medical field, or seismic differences in earthquakes, among many other fascinating topics.

The Derivative of e^x

Interestingly, the slope of a line tangent to the graph of $y = e^x$ at any point (x_1, y_1) is always equal to the y-coordinate at that point. Try it yourself and see. Use your graphing calculator to graph $y = e^x$. Trace to any point on the graph, and make note of the y-coordinate. Now use the "nDeriv(" feature in the Calc menu, and check the derivative of the function at your point. It should be the same as your y-coordinate. This means that the derivative of e^x is actually e^x!

Of course, as you have already learned, if the function is more complicated, using the chain rule may be necessary to find the correct derivative. For instance, if you want the derivative of $y = e^{3x}$, you can think of it as a composite of $y = e^u$ and $u = 3x$. The derivative is computed as follows:

$$\begin{aligned}\frac{dy}{dx} &= \frac{dy}{du} \cdot \frac{du}{dx} \\ &= e^u \cdot 3 \\ &= 3e^{3x}\end{aligned}$$

Taking the derivative of the various powers of e tends to be one of the easier tasks in calculus. Simply rewrite the entire exponential function and multiply it by the derivative of the exponent. The difficulty of the problem really depends on how hard it is to take the derivative of the exponent.

If u is any function of x, and $y = e^u$, then $\frac{dy}{dx} = e^u \frac{du}{dx}$. Note that the derivative is simply the original function times the derivative of the exponent.

The Derivative of ln(x)

Taking the derivative of the natural log function is also fairly straightforward. Studying a graph of the function and remembering that a derivative gives the slope of the graph at any point should help you understand the derivative rule intuitively. Figure 8-1 shows the graph of $y = \ln(x)$ with the tangent line drawn at $(1, 0)$.

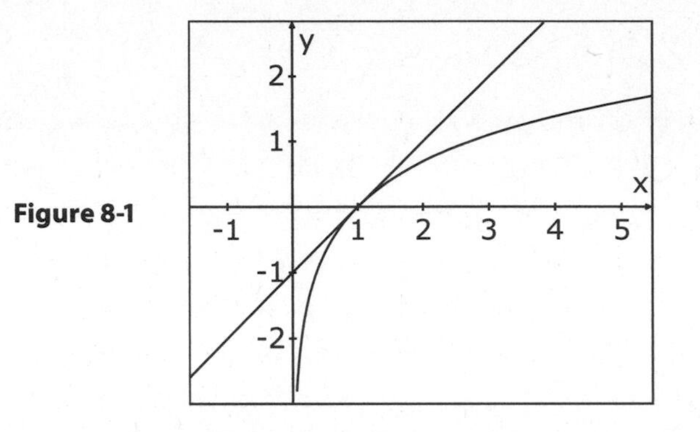

Figure 8-1

What would happen to the slope of the line if the x-coordinate for the point of tangency were increased? The graph of $y = \ln(x)$ gets less steep as you move to the right, so the value of the derivative must be decreasing. The specific relationship is that the value of the derivative on $y = \ln(x)$ is always the reciprocal of the x-coordinate. One way to write this in symbols is $\frac{d}{dx}[\ln(x)] = \frac{1}{x}$.

As with e^x, finding the derivative of the natural logarithm function can also require the chain rule. If the argument of the natural logarithm is any function other than just x, then you multiply the reciprocal of the argument by the derivative of the argument. If $f(x) = \ln(x^2 + 4)$, the argument is $x^2 + 4$. To get the derivative of $f(x)$, write the reciprocal of the argument, and multiply it by the derivative of the argument.

$$\begin{aligned} f'(x) &= \frac{1}{x^2+4} \cdot \frac{d}{dx}(x^2+4) \\ &= \frac{1}{x^2+4} \cdot 2x \end{aligned}$$

If u is any function of x, and $y = \ln(u)$, then $\frac{dy}{dx} = \frac{1}{u} \cdot \frac{du}{dx}$. Note that the derivative is simply the reciprocal of the argument times the derivative of the argument.

The Derivative of $\log_a x$

Succeeding with calculus requires recalling some college algebra, and among the things you need to know are the properties of logarithms. Such knowledge is helpful in multiple-choice testing situations, where the answer choices may not exactly match the result of your work. Knowing these properties can also help you understand where a particular derivative comes from, rather than having to blindly memorize a derivative formula.

This is the case with the derivative of $\log_a x$, which is read "log base a of x." Do you remember a property called the Change of Base Theorem? It enables you to rewrite a logarithm of any base as a ratio of logarithms with a new base. The formula is $\log_a x = \frac{\log_c x}{\log_c a}$, where c is any real number. When you are finding a derivative, this can be very useful if you choose c to be e, the base for natural logarithms. When this is done, $\log_a x$ becomes $\frac{\ln(x)}{\ln(a)}$. When you rewrite the original logarithm in this form, because $\ln(a)$ is a constant, all you have to do to take the derivative is use the rule for the derivative of $\ln(x)$. For instance, if $y = \log_3 x$, rewrite it as $y = \frac{\ln(x)}{\ln(3)}$. Now when you take the derivative, the $\ln(3)$ just sits there, so $\frac{dy}{dx} = \frac{\frac{1}{x}}{\ln(3)}$, or, cleaned up a bit, $\frac{1}{x \cdot \ln(3)}$.

For more complicated logarithms, apply the chain rule when necessary. For example, if $y = \log_9(5x+1)$, first use the Change of Base Theorem to get $y = \frac{\ln(5x+1)}{\ln(9)}$. Then simply take the derivative of the numerator so $\frac{dy}{dx} = \frac{\frac{1}{5x+1} \cdot 5}{\ln(9)}$. Remember that the extra 5 comes from the derivative of $5x+1$.

Logarithmic Differentiation

Even though the phrase *logarithmic differentiation* sounds intimidating, it's really just a four-step method to help take derivatives and to prove another derivative rule. The process uses another one of those long-forgotten logarithm rules. This time the algebra rule to remember is $\ln(x^n) = n \cdot \ln(x)$. In words, the exponent on an argument of a logarithm can be "rolled down" in front of the logarithm as a multiplier. You have already learned how to take the derivative of a power function, a variable raised to a number. You will soon learn a rule for taking the derivative of any exponential function, a number raised to a variable. But there is no short or simple rule for finding the derivative of a variable raised to a variable, so the process of logarithmic differentiation has to be used.

Suppose you wanted to find the slope of $y = x^{\cos(x)}$ at $x = 1$. You will need to evaluate the derivative when $x = 1$. To find the derivative, if you apply the power rule by bringing $\cos(x)$ down in front and reducing the exponent by 1, you get $\frac{dy}{dx} = \cos(x) \cdot x^{[\cos(x)-1]}$, the most common *wrong* answer!

Perhaps the most commonly used derivative rule is the power rule, $\frac{d}{dx}(u^n) = n \cdot u^{(n-1)} \frac{du}{dx}$, where *u* is any function of *x*, and *n* is a real number. But this rule can be used *only* on power functions. For power functions, the base has to be a variable expression, and the exponent must be a number.

Thankfully, logarithmic differentiation can help you change the form of what you are differentiating so that you can use different rules. You will end up using implicit differentiation on the left-hand side of the equation and the product rule on the right-hand side. The first step is to take the natural logarithm of both sides of the equation and roll the exponent down in front as a multiplier. When you do this, $y = x^{\cos(x)}$ becomes $\ln(y) = \ln\left(x^{\cos(x)}\right)$, which can be rewritten as $\ln(y) = \cos(x) \cdot \ln(x)$. The second step is to differentiate *both* sides of the equation. Use the derivative of a natural logarithm on the left-hand side of the equation and the product rule on the right-hand side of the equation, because the right-hand side is a product of two functions of x. One way to write your intention to do this is $\frac{d}{dx}[\ln(y)] = \frac{d}{dx}[\cos(x)\cdot\ln(x)]$.

Did you get $\frac{1}{y}\cdot\frac{dy}{dx} = \cos(x)\cdot\frac{1}{x} + \ln(x)\cdot[-\sin(x)]$? It may be messy, but it is correct. Now all you have to do is multiply both sides of the equation by y to get $\frac{dy}{dx}$ by itself: $\frac{dy}{dx} = y\left[\frac{\cos(x)}{x} - \ln(x)\cdot\sin(x)\right]$. The final step is to replace y with the original equation, $y = x^{\cos(x)}$, so that the entire derivative expression is in terms of x. Your derivative expression is now $\frac{dy}{dx} = x^{\cos(x)}\left[\frac{\cos(x)}{x} - \ln(x)\cdot\sin(x)\right]$.

Of course, all good calculus students make sure they have answered the original question. To get the slope of the graph $y = x^{\cos(x)}$ at $x = 1$, it is necessary to evaluate the derivative by substituting 1 in for x: $\left.\frac{dy}{dx}\right|_{x=1} = 1^{\cos(1)}\left[\frac{\cos(1)}{1} - \ln(1)\cdot\sin(1)\right]$. Do you remember that $\ln(1) = 0$?

$$= \cos(1)$$

To summarize, you get the derivative of a function raised to a function by using the four steps of logarithmic differentiation: (1) Take the natural logarithm of both sides of the equation, and roll the exponent down in front. (2) Differentiate both sides of the equation. (3) Multiply both sides of the equation by y. (4) Replace y with its function of x so that the final derivative is all in terms of x.

The Derivative of a^x

Exploring on the graphing calculator, you intuitively learned that the derivative of the exponential function e^x is itself, e^x, but what if the base is something other than e? Using logarithmic differentiation, you can now prove what you accepted from exploration, and you can figure out what happens when the base is not e.

Start with $y = e^x$ and take the natural logarithm of both sides of the equation to get $\ln(y) = \ln(e^x)$. Next, roll the exponent down in front of the logarithm so that $\ln(y) = x\ln(e)$. Now remember that the natural logarithm function is the inverse of e^x, so conveniently $\ln(e) = 1$, and your equation becomes $\ln(y) = x$. If you take the derivative of both sides of the equation, $\frac{d}{dx}[\ln(y)] = \frac{d}{dx}[x]$, you get $\frac{1}{y} \cdot \frac{dy}{dx} = 1$, or $\frac{dy}{dx} = y$. Simply substitute $y = e^x$ on the right-hand side of the equation, and you get $\frac{dy}{dx} = e^x$. The first task is accomplished! You should now be convinced that the derivative of e^x is indeed e^x.

To answer the question asked a moment ago, try the same method with any base other than e, perhaps 5. If $y = 5^x$, then $\ln(y) = \ln(5^x)$, and $\ln(y) = x\ln(5)$. You may already have noticed that $\ln(5)$ does not disappear nicely, as $\ln(e)$ did. But $\ln(5)$ is just a constant, so taking the derivative of $x\ln(5)$ is no more difficult than taking the derivative of $7x$. Differentiate both sides of $\ln(y) = x\ln(5)$, and you get $\frac{1}{y} \cdot \frac{dy}{dx} = \ln(5)$. This transforms to $\frac{dy}{dx} = y \cdot \ln(5)$ and, after substitution, becomes $\frac{dy}{dx} = 5^x \cdot \ln(5)$. If you notice, the only difference in the result is that the natural logarithm of the base becomes 1 when the base is e and does not simplify when the base is not e. To generalize, if $y = a^x$, where a is any positive number, then $\frac{dy}{dx} = a^x \cdot \ln(a)$.

Once you see and understand the pattern, you do not have to do logarithmic differentiation every time. When the exponent is just x, to take the derivative you simply rewrite the function and multiply by the natural logarithm of the base. If $f(x) = 9^x$, then $f'(x) = 9^x \cdot \ln(9)$. But all good calculus students remember the chain rule, so if the exponent is more complicated, you should handle the problem just like the other derivatives in this chapter. If $y = 2^{(x^3)}$, then let $y = 2^u$ and $u = x^3$.

Use $\frac{dy}{dx}=\frac{dy}{du}\cdot\frac{du}{dx}$. $\frac{dy}{du}=2^u\cdot\ln(2)$ and $\frac{du}{dx}=3x^2$. Multiplying, you get $\frac{dy}{dx}=2^u\ln(2)\cdot 3x^2=2^{(x^3)}\ln(2)\cdot 3x^2$.

Skill Check

It's time for you to tackle a few problems on your own to see how well you are learning these new derivatives.

1. Find the derivative of $y=3^{(x^2-x)}$.

2. Find the equation of the line tangent to the graph of $y=\ln(x^3-7)$ at the point (2,0).

3. Find the instantaneous rate of change of $f(x)=x^{(x+2)}$ when $x=1$.

4. If $y=\log_5(u)$ and $u=1+\sqrt{x}$, find $\frac{dy}{dx}$.

5. You recently invested $10,000 at 4% interest compounded quarterly; therefore, your balance is modeled by $A=10000(1.01)^{4t}$, where t is the number of years the money remains invested. Determine how rapidly your investment is growing 3 years after the initial investment. (For this problem, it's okay to use a calculator.)

CHAPTER 9

Derivatives of Inverse Trigonometric Functions

In all fields of mathematics, the concept of an inverse is vitally important. The simple step of solving a single-variable equation by subtracting a constant from both sides of the equation is using the additive inverse. When you divide both sides of an equation by the coefficient of the variable, you are using the multiplicative inverse. This chapter examines the inverse trigonometric functions and their rates of change.

Inverse Functions

As you may recall, a function is a set of ordered pairs where each element of the domain is paired with only one element of the range. In other words, each x value produces only one y value. Whether a graph represents a function is often investigated by a vertical-line test. If there is no vertical line that can intersect a graph in more than one point, it passes the vertical-line test and is a function. No x value is paired with more than one y value. There is also a more specific class of functions called *one-to-one functions* for which each x is paired with only one y and each y is paired with only one x. A simple parabola is a function, but not a one-to-one function. This is because two x values that are equal distances on either side of the axis of symmetry produce the same y value.

Recall also that, given an equation, its inverse is found by replacing each x with a y, replacing each y with an x, and, if possible, isolating y again. For example, to get the inverse of $y = x^2 + 2$, you must write $x = y^2 + 2$ and isolate y again. The geometric result of interchanging x and y in an equation is reflecting the graph of the original function over the line $y = x$. Another result of interchanging x and y is that the domains and ranges of the inverses are switched. The domain of the original function becomes the range of the inverse. Figure 9-1 shows the graphs of $y = x^2 + 2$, its inverse $x = y^2 + 2$, and the line $y = x$ on the same grid.

Figure 9-1

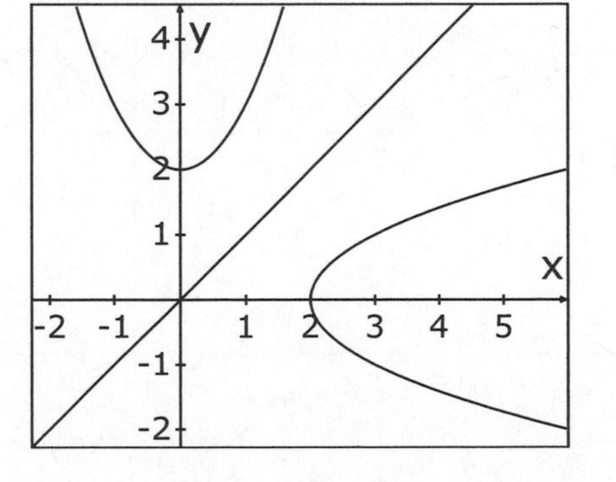

If a function such as the parabola $y = x^2 + 2$ is not one-to-one, then its inverse will not be a function. Note in Figure 9-1 that the graph of the inverse does not pass the vertical-line test. With the exception of the vertex, every x value is paired with two y values. In this case, a more accurate name is *inverse relation*, because the inverse is not a function. It is most advantageous to work with one-to-one functions, because the inverse of a one-to-one function is also guaranteed to be a function. This eliminates much confusion and ambiguity in responding to questions about the function and its inverse. It also ensures consistency when you are determining and discussing properties of those functions.

No trigonometric functions are one-to-one. As a result, their inverses are relations but not functions. Mathematicians dealt with this issue by limiting the domain of the original function to make it one-to-one, thereby making its inverse a function. For instance, the shortest subset of the domain of $y = \sin(x)$, which produces the entire range of $-1 \leq y \leq 1$, is $\left[\frac{-\pi}{2}, \frac{\pi}{2}\right]$. Therefore, the inverse function of the sine has a domain of $-1 \leq x \leq 1$ and a range of $\left[\frac{-\pi}{2}, \frac{\pi}{2}\right]$. Figure 9-2 shows multiple periods of a sine graph and its inverse relation. Figure 9-3 shows the sine graph with limited domain and its inverse function.

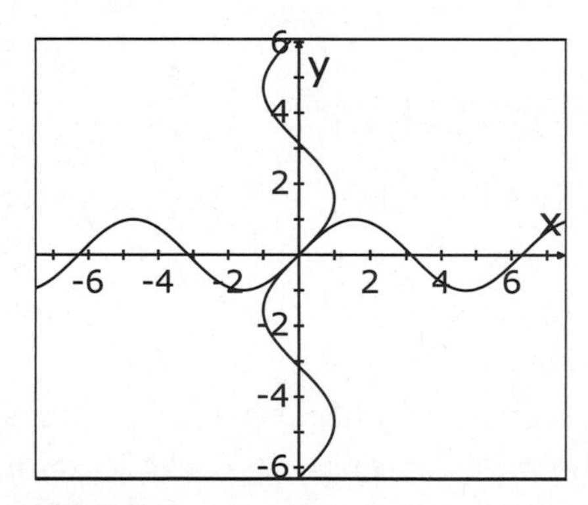

Figure 9-2

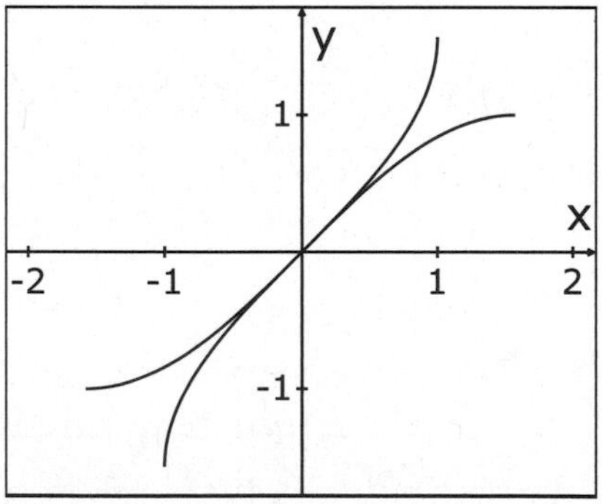

Figure 9-3

You must consider the principal range whenever you evaluate an inverse trigonometric function. For instance, there are an infinite number of radian measures for which the sine is $\frac{\sqrt{3}}{2}$, but the inverse sine function, $\sin^{-1}\left(\frac{\sqrt{3}}{2}\right)$, equals only $\frac{\pi}{3}$ because of the principal range. Be sure to review the principal ranges of the inverse trigonometric functions.

Notation

A quick word about the notation for inverse functions—and especially inverse trigonometric functions—is necessary. Unfortunately, a notation that looks just like a power of negative one is most commonly used, so the proper interpretation of its meaning must come from the context of the problem. In general function notation, $f(x)$ and $f^{-1}(x)$ are more often inverses than reciprocals. With trigonometric functions, the inverse can be indicated with the same notation or with the prefix "arc." Thus both $\sin^{-1}(x)$ and $\arcsin(x)$ mean the inverse of the sine function. Again, context must be carefully considered.

Derivatives of Inverse Functions

In calculus, a natural point of interest about inverse functions is how their derivatives are related. The deepest understanding is best achieved by studying the relationship analytically and graphically. The chain rule also plays a major role in shedding light on the topic.

If two functions are inverses, then the output of their composition will always be equal to the initial input. In symbols, if f and g are one-to-one inverse functions, then $f(g(x)) = x$ and $g(f(x)) = x$. The output of the composition of functions f and g is the same as the input, x. Example 9-1 uses the chain rule to shed light on the relationship between the inverse functions f and g.

EXAMPLE 9-1

Given that f and g are one-to-one inverse functions, differentiate $g(f(x)) = x$ to find a relationship between their derivatives.

$$\frac{d}{dx}[g(f(x))] = \frac{d}{dx}(x)$$

$$g'(f(x)) \cdot f'(x) = 1$$

$$f'(x) = \frac{1}{g'(f(x))}$$

The result achieved in Example 9-1 confirms that the derivatives of inverse functions are reciprocals of one another *at the correct points*. It is critical to understand that the derivatives are not being evaluated at the same x value. The derivative of function g is being evaluated at its x value, which is the corresponding y value from function f. This is because the range of f is the domain of g. Figure 9-4 provides a graphical look at this relationship. Inverse one-to-one functions are shown with tangent lines at the corresponding points of reflection over the line $y = x$. The slopes of the tangent lines are reciprocals; but on one function, x is the independent variable, and on the other function, $f(x)$ is the independent variable.

Figure 9-4

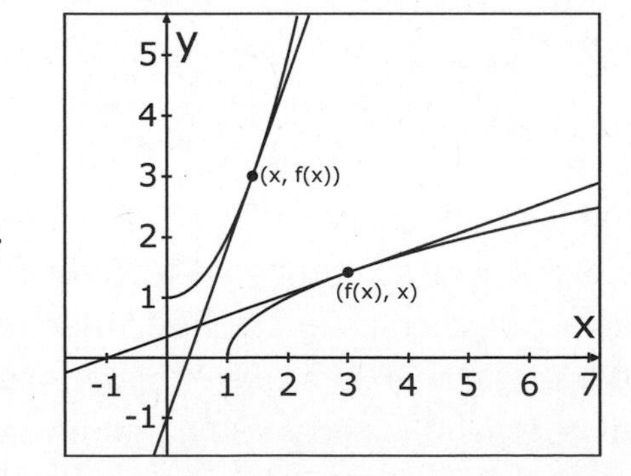

Given one-to-one inverse functions *f* and *g*. Let P(*x*, *f*(*x*)) be a point on *f*, and let Q(*f*(*x*),*x*) be the point on *g* that is the reflection of *P* across the line $y = x$. Then the derivative of *f* at *P*, evaluated at its abscissa, is the reciprocal of the derivative of *g* at *Q*, evaluated at its abscissa. In symbols, $f'(x) = \frac{1}{g'(f(x))}$.

Example 9-2 shows how this rule applies to specifically defined inverse functions.

EXAMPLE 9-2

The functions $f(x) = x^3$ and $g(x) = \sqrt[3]{x}$ are inverses. $f(2) = 8$ and $g(8) = 2$ are corresponding points of reflection. Show that the derivative of *f* at (2,8) is the reciprocal of the derivative of *g* at (8,2).

$f'(x) = 3x^2$ and $g'(x) = \frac{1}{3}x^{\left(\frac{-2}{3}\right)}$ Differentiate each function.

$f'(2) = 3 \cdot 2^2 = 12$

$g'(8) = \frac{1}{3} \cdot 8^{\left(\frac{-2}{3}\right)} = \frac{1}{12}$

Clearly $f'(2)$ and $g'(8)$ are reciprocals.

Derivative of the Inverse Sine Function

All six trigonometric functions have inverses with carefully defined domains and ranges to ensure that the inverses are functions. Of course, they also have derivatives that are best established using implicit differentiation and some well thought-out right triangle trigonometry. For your overview, justifying all six would be a bit tedious, so inverse sine and inverse tangent will be established, and the remaining four will be provided without proof.

Consider the function $y = \sin(x)$. The first step in determining its inverse is to interchange the x and y variables. Thus one form of the inverse of the sine function is $x = \sin(y)$, which is equivalent to $y = \sin^{-1}(x)$. The former is much more useful than the latter for discovering a form of the derivative of the inverse sine function. Before you differentiate implicitly, though, Figure 9-5 provides an important visual understanding of the relationship $x = \sin(y)$. Note that y is the angle, and x is the opposite side in a right triangle with hypotenuse length 1. Therefore, by right triangle relationships, $\sin(y) = x$. Additionally, the Pythagorean Theorem is used to determine the length of the second leg.

Figure 9-5

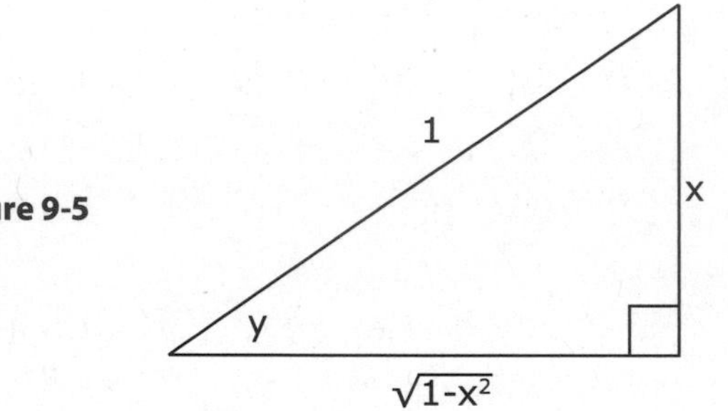

Because $x = \sin(y)$ is the inverse of the sine function, finding $\frac{dy}{dx}$ from it provides the derivative of the inverse sine function. Differentiate both sides of the equation with respect to x, and use implicit differentiation and the chain rule on the right-hand side of the equation. You get $1 = \cos(y)\frac{dy}{dx}$, which means $\frac{dy}{dx} = \frac{1}{\cos(y)}$. This result is more useful if it is dependent on x, so now substitute for $\cos(y)$ from Figure 9-5. Using the adjacent side divided by the hypotenuse yields $\cos(y) = \sqrt{1 - x^2}$. The conclusion is that if $y = \sin^{-1}(x)$, then $\frac{dy}{dx} = \frac{1}{\sqrt{1 - x^2}}$.

Derivatives of the Other Inverse Functions

Establishing the derivative of the inverse tangent function is done similarly, but it requires a drawing different from Figure 9-5. If $y = \tan(x)$, then one form of the inverse function is $x = \tan(y)$. This means that in the right triangle, the side opposite the angle must be x, and the side adjacent must be 1. Figure 9-6 shows the right triangle with all three sides labeled.

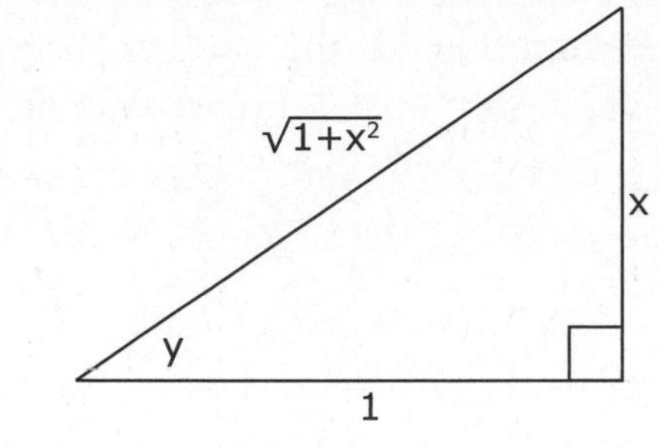

Figure 9-6

Differentiating $x = \tan(y)$ with respect to x yields $1 = \sec^2(y)\frac{dy}{dx}$.

Isolating $\frac{dy}{dx}$ leads to $\frac{dy}{dx} = \frac{1}{\sec^2(y)}$. From Figure 9-6, $\sec(y) = \frac{\sqrt{1+x^2}}{1}$, so $\sec^2(y) = 1 + x^2$. By substitution, if $y = \tan^{-1}(x)$, then $\frac{dy}{dx} = \frac{1}{1+x^2}$.

The remaining four derivatives can be derived by similar means. If you want a challenge, try deriving one of them on your own. The following table summarizes all the derivatives of all six inverse trigonometric functions.

Function	Derivative
$y = \sin^{-1}(x)$	$\frac{dy}{dx} = \frac{1}{\sqrt{1-x^2}}$
$y = \cos^{-1}(x)$	$\frac{dy}{dx} = \frac{-1}{\sqrt{1-x^2}}$
$y = \tan^{-1}(x)$	$\frac{dy}{dx} = \frac{1}{1+x^2}$
$y = \cot^{-1}(x)$	$\frac{dy}{dx} = \frac{-1}{1+x^2}$
$y = \sec^{-1}(x)$	$\frac{dy}{dx} = \frac{1}{\lvert x\rvert\sqrt{x^2-1}}$
$y = \csc^{-1}(x)$	$\frac{dy}{dx} = \frac{-1}{\lvert x\rvert\sqrt{x^2-1}}$

Even though this seems like a lot of information, there are really only three derivatives to remember. Do you notice that the derivative of any function and its cofunction differ only in sign? Each of the cofunctions is simply the opposite of its normal function. Seeing patterns like this is very useful for organizing the large amount of factual information in calculus.

The Chain Rule Again!

The given derivative rules, as given, used just a simple x as the input to the inverse trigonometric function. As with all the other derivative rules, if the input has a derivative of anything other than 1, then the chain rule multiplier must be included. Example 9-3 uses Leibniz notation to show how this is handled.

EXAMPLE 9-3

If $y = \tan^{-1}(x^3 + 5)$, find $\dfrac{dy}{dx}$.

Let $y = \tan^{-1}(u)$ and $u = x^3 + 5$ — Decompose the function.

$\dfrac{dy}{du} = \dfrac{1}{1+u^2}$ and $\dfrac{du}{dx} = 3x^2$ — Differentiate using the pattern for inverse tangent.

$$\frac{dy}{dx} = \frac{dy}{du} \cdot \frac{du}{dx}$$

$\dfrac{dy}{dx} = \dfrac{1}{1+u^2} \cdot 3x^2$ — Substitute for $\dfrac{dy}{du}$ and $\dfrac{du}{dx}$.

$\dfrac{dy}{dx} = \dfrac{1}{1+(x^3+5)^2} \cdot 3x^2$ — Substitute for u.

Skill Check

1. Find $\dfrac{dy}{dx}$ if $y = \cos^{-1}(3x)$.
2. Find $f'(1)$ if $f(x) = x^4 \tan^{-1}(x)$.
3. Find the equation of the line tangent to $y = \sec^{-1}(x)$ at $x = 2$.
4. If $y = \cot^{-1}\left(\dfrac{1}{x}\right)$, find $\dfrac{dy}{dx}$.
5. Find $g'(x)$ if $g(x) = \csc^{-1}(e^{5x})$.

CHAPTER 10

Higher-Order Derivatives

Presumably, by this point you understand a derivative as a rate of change of a function and have developed skill at finding derivatives for a wide variety of functions. This next step in your journey through calculus takes you into a vast assortment of applications of the derivative and reveals that there is meaning in the derivative of the derivative—and beyond.

Notation

As with all topics in calculus (or in any course of study, for that matter), becoming familiar with vocabulary and notation increases your ability to learn and enjoy what you are exploring. This is the case as you move beyond finding a single derivative.

As you have seen and experienced, the derivative of a given function is itself another function. Thus there is no reason why you cannot take the derivative of the result of a first derivative. Doing so produces the second derivative. Repeat the process again on the second-derivative result, and you have taken the third derivative. As long as no derivative along the way ended up as a constant, whose next derivative is 0, you could continue taking derivatives of the same function forever! Believe it or not, there are practical uses for that.

Of course, vocabulary and conventions of notation need to be established. The *order* of a derivative is the name given to indicate which derivative number it is. Thus the third derivative of a function is called a third-order derivative, the fourth derivative is a fourth-order derivative, and so on.

For the most part, the notation is a natural extension of what you have already experienced, but there are a few odd twists. The second derivative of a function can be denoted $f''(x)$, which is read "f double prime of x." It can also be written as $y''(x)$ or just as y''. The somewhat odd notation is the Leibniz notation $\frac{d^2y}{dx^2}$, which is read "d squared y, dx squared." For derivatives of higher order, the notations can just be extended until the use of prime notation begins to get cumbersome. The prime marks are then replaced with either digits in parentheses or roman numerals in parentheses. For instance, the eighth derivative of a function could be noted as either $\frac{d^8y}{dx^8}$ or $f^{(8)}(x)$.

What the Second Derivative Means

It is natural to wonder whether the second derivative has any practical use. The answer to that is a resounding "Yes!" It has practical application in a lot of different fields.

For functions in general, if the first derivative measures the rate of change of the function, then the second derivative measures the rate of change of the first derivative. In other words, it provides information about the change in the slope of a graph. That knowledge can help you achieve a more accurate understanding of the shape and behavior of a graph.

Economists are not only interested in trends in the economy indicated by changes in key economic factors; they are also interested in how rapidly those changes are taking place—that is, in the rate of change of the rate of change in those factors. The question is not just "Is the gross national product increasing?" but "Is the gross national product increasing more rapidly or is it increasing more slowly?"

One of the many real-world applications of first and second derivatives is in the field of UV-visible spectrophotometry. They are used in the analysis of the intensity of wavelengths in various parts of the spectrum. Suppose a doctor prescribes a certain medicine for a patient who has a fever. It may take some time for the medicine to begin working. The doctor certainly wants to know whether the patient's temperature is rising at an increasing rate or a decreasing rate. The temperature rising at an increasing rate might indicate that the medicine is not working the way the doctor had hoped.

Implications for Particle Motion

One of the most important applications of the second derivative is related to motion. If the position of a moving object is dependent on time, then it may be possible to describe that position with a function. Suppose that an object's distance from a fixed point is being measured in meters and that the position is dependent on time measured in minutes. The first derivative measures the rate of change of position with respect to time. This derivative provides the instantaneous velocity of the object in units of meters per minute. The second derivative measures the rate of change of velocity with respect to time, so the units are meters per minute per minute. If you have had any physics experience, you know that this second derivative describes acceleration. Example 10-1 offers a first look at the second derivative of a position function.

EXAMPLE 10-1

The distance a freely falling object travels is described by the equation $s(t) = 4.9t^2$, where s is measured in meters and t is measured in seconds. Find the acceleration of the object as it falls. Include units in your answer.

$v(t) = s'(t) = 9.8t$ meters per second. Differentiate position to get velocity.

$a(t) = v'(t) = s''(t) = 9.8$ meters per second per second. Differentiate velocity to get acceleration.

If the number 9.8 looks familiar, that is because it is the acceleration, on Earth, due to gravity. There is frequently a negative sign associated with it, but the role of the sign is simply to indicate a predetermined direction. Traditionally, downward motion is described as having negative velocity and negative acceleration when the position function is defined as height above the ground. In Example 10-1, the position function was defined as distance the object had fallen, essentially reversing the standard.

Higher Derivatives of Explicit Functions

As you may anticipate, the actual process of finding a higher derivative of an explicitly defined function is relatively straightforward, but the results can quickly get rather complicated. You need to be alert to multiple uses of the product rule, quotient rule, or chain rule. Beyond the simple mechanics of finding higher-order derivatives, sometimes problems are approached numerically or through the use of pattern discovery. As previously stated, pattern recognition in mathematics is an important and powerful skill. The examples that follow will take you through an increasing progression of finding and understanding higher derivatives.

EXAMPLE 10-2

Given $y = x^3 + 5x^2 - 8x$, find $\frac{d^2y}{dx^2}$.

$$\frac{dy}{dx} = 3x^2 + 10x - 8$$

$$\frac{d^2y}{dx^2} = 6x + 10$$

Example 10-2 was easy, wasn't it? Simply apply the power rule twice. Even though it will be left unfinished, Example 10-3 demonstrates how a second derivative can quickly get very messy. The point is made without doing all the work!

EXAMPLE 10-3

If $f(x) = x^3 \cdot \sqrt{x^2 - 6x}$, find $f''(x)$.

$f'(x) = x^3 \cdot \frac{1}{2}(x^2 - 6x)^{\frac{-1}{2}}(2x - 6) + 3x^2 \cdot \sqrt{x^2 - 6x}$ Product rule, power rule, and chain rule were all used in the first step.

$f'(x) = \frac{x^4 - 3x^3}{\sqrt{x^2 - 6x}} + 3x^2 \cdot \sqrt{x^2 - 6x}$ Simplify the first derivative.

At this point there are two options. The first is to do more algebra to combine the terms of the first derivative. That may actually be the more prudent step. The second option is to use the quotient rule and chain rule on the first term and the product rule and chain rule on the second term.

Although mechanically involved and rigorous problems do arise from time to time in any mathematical study, the study of calculus does not have to be characterized by such problems. The concepts of the course can be learned just as well—and sometimes more effectively—in the context of algebraically manageable examples or even problems that involve no symbolic manipulation at all. Examples 10-4 and 10-5 demonstrate this idea.

EXAMPLE 10-4

Consider the following statement: "The cost of a college education is increasing at an increasing rate." Let $c(t)$ represent the cost of a college education, where t is the number of years after 2000. Is the second derivative of $c(t)$ positive or negative? Explain.

The second derivative of $c(t)$ is positive. The first part of the sentence, "The cost of a college education is increasing," indicates that $c'(t)$ is positive, because $c(t)$ has a positive slope. The second part of the sentence, "at an increasing rate," means that $c'(t)$ is increasing. Therefore, the slope of $c'(t)$ is positive, which means $c''(t)$ is positive.

EXAMPLE 10-5

Let function $h(t)$ be a monotonically increasing, twice-differentiable function on the domain $[0,4]$ with the values shown in the following table. Use the values in the table to explain whether $h''(t)$ is positive or negative.

t	$h(t)$
0	3
1	8
2	12
3	14
4	15

On the given domain, $h'(t)$ can be estimated in each interval by the change in h divided by the change in t. From 0 to 1, $h'(t)$ is about 5; from 1 to 2, $h'(t)$ is about 4; from 2 to 3, $h'(t)$ is about 3; and from 3 to 4, $h'(t)$ is about 1. Therefore, as t increases, $h'(t)$ is decreasing, so $h''(t)$ must be negative.

If a function were fit to the United States national debt, plotting debt as a function of the number of years after 1900, would its second derivative be positive or negative?

Unfortunately, the second derivative would be positive, because the national debt is increasing at an increasing rate.

The next example demonstrates that pattern finding can be a useful tool and a significant time saver in certain situations.

EXAMPLE 10-6

Given $f(x) = -\ln(x)$. Find a general expression for the nth derivative of $f(x)$.

$$f'(x) = -\frac{1}{x} = -x^{-1}$$

$$f''(x) = 1x^{-2}$$

$$f'''(x) = -2x^{-3}$$

$$f^{(4)}(x) = 6x^{-4}$$

$$f^{(5)}(x) = -24x^{-5}$$

Note that the sign in front alternates between positive and negative and is positive on the even powers of x. Note also that the exponent on the result equals the order of the derivative. The hardest pattern to see is the value of the coefficient. You must recognize that each time you apply the power rule, the previous derivative gets multiplied by the exponent being brought down in front. This is a factorial pattern. Ignoring the sign changes, $2! = 2$ and $3! = 6$ and $4! = 24$. On the fifth-order derivative, the coefficient is four factorial, or 4!, so the factorial is always 1 less than the order. The general nth derivative is therefore $f^{(n)}(x) = (-1)^n (n-1)!(x)^n$.

Second Derivatives of Implicit Functions

Previously, you learned to find $\frac{dy}{dx}$ for implicit functions. In certain situations you will need to find a second derivative implicitly. Rarely is a third- or higher-order derivative needed for an implicit function. Even though a second derivative can get mechanically messy rather quickly, the actual process involves only one step more than finding the first derivative. Most of the time, the first derivative of an implicit function will be another implicit function. Just differentiate the result implicitly, as you did to find the first derivative. If the first derivative has a y anywhere in the expression, then $\frac{dy}{dx}$ will appear in the second derivative. The one extra step is to replace $\frac{dy}{dx}$ everywhere it appears in your second-derivative expression. Example 10-7 demonstrates this, starting with the first derivative already having been determined.

EXAMPLE 10-7

If $\frac{dy}{dx} = \frac{2y+1}{5x-3}$, find the second derivative, $\frac{d^2y}{dx^2}$.

$$\frac{d}{dx}\left(\frac{dy}{dx}\right) = \frac{d}{dx}\left(\frac{2y+1}{5x-3}\right)$$

On both sides of the equation, prepare to find the derivative with respect to x.

$$\frac{d^2y}{dx^2} = \frac{(5x-3)\frac{d}{dx}(2y+1) - (2y+1)\frac{d}{dx}(5x-3)}{(5x-3)^2}$$

Use quotient rule on the right-hand side.

$$\frac{d^2y}{dx^2} = \frac{(5x-3)\cdot 2\frac{dy}{dx} - (2y+1)\cdot 5}{(5x-3)^2}$$

Differentiating the y term produces a $\frac{dy}{dx}$.

$$\frac{d^2y}{dx^2} = \frac{(5x-3)\cdot 2\left(\frac{2y+1}{5x-3}\right) - (2y+1)\cdot 5}{(5x-3)^2}$$

Replace $\frac{dy}{dx}$ with $\frac{2y+1}{5x-3}$.

$$\frac{d^2y}{dx^2} = \frac{2(2y+1)-(2y+1)\cdot 5}{(5x-3)^2}$$ Simplify the numerator.

$$\frac{d^2y}{dx^2} = \frac{-6y-3}{(5x-3)^2}$$ Combine like terms.

As is the case for explicit functions, even if the graph of an implicit relation is not a function, the second derivative provides information about rate of change of the first derivative and about the shape of the graph. This concept will be explored further in the next chapter.

Before forging ahead with any higher-order derivative, each step of the way you should check to see whether the current expression will simplify and make taking the next derivative much easier. You can often save yourself a lot of algebraic headaches by doing this.

Skill Check

Do the best you can to answer these Skill Check questions without looking at the solutions first. You may still be at a point where you need to look up the derivatives of some of the functions presented in earlier chapters. Don't worry about that. Go ahead and look them up. As you use the derivatives more and more, you will find it increasingly easy to recall the facts you need.

1. If $k(x)=3x^4+x^2-3x$, find the ratio of $k'(x)$ to $k''(x)$ at the origin.
2. Is the derivative of $h(x)=e^{-2x}$ increasing or decreasing at $x=3$? Show work to support your answer.
3. A model rocket is shot into the air. While the engine is still firing, its height above the ground is modeled by the function $s(t)=-4t^3+12t^2+72t$, where s is measured in feet and t is measured in seconds. Find the velocity and acceleration of the rocket at $t=2$ seconds. Is the engine still firing at that moment?

4. The mass of the polar ice caps can be modeled as a function of time, call it $m(t)$. If scientists discover that the polar ice caps are still melting more and more each year, what are the signs of the first and second derivatives of $m(t)$?
5. Find the second derivative if $\frac{dy}{dx} = \frac{x^5}{y^2 + 1}$.

CHAPTER 11

Graph Analysis Using Derivatives

In this day and age of graphing calculators, computers, and numerous other hand-held devices that quickly and easily generate graphs, producing a graph by hand may seem to be obsolete. But the real value in curve sketching in calculus is the connection between a graph and its derivatives and what each one tells you about the other.

Curve Sketching

In prior courses and in the early parts of your study of calculus, you have already begun learning about producing and interpreting graphs. One of the expected prerequisites for studying calculus is to be familiar with the graphs of the nine basic parent functions. Your previous studies of changes to the parent function equations and of the transformations caused by those changes have expanded your skills to sketch a larger variety of graphs. For instance, knowing that $y=(x+8)^2$ simply shifts the graph of $y=x^2$ to the left 8 units makes it easy to produce a quick sketch. Examining limits and what they tell you about vertical asymptotes, horizontal asymptotes, and removable discontinuities in a graph is another useful tool for curve sketching.

What characteristics of a graph cause the derivative of the function not to exist?

The derivative will not exist if a graph has a corner, cusp, removable discontinuity, vertical asymptote, or vertical tangent at a certain point.

The next step is to be able to produce a graph of the first and second derivatives from a given graph. An even more advanced skill is to determine what the graph of a function looks like, given only information about its first and second derivatives. Often, the only knowledge a researcher has is information about how something is changing at certain moments in time. It is very useful to be able to build a potential model from that information in order to interpolate or extrapolate data to make predictions and informed forecasts.

Producing a Graph of a First Derivative

One thing to remember as you begin to produce graphs of derivatives, and then graphs from derivatives, is that producing exact graphs is not the initial goal. The initial goal has more to do with general shape and graphical behavior. Much of the graphing will be done without scale. Work at developing a strong understanding of the graphical connections between a function and its first two derivatives.

Recall that a derivative is the slope of a function at any given point. As a result, when you look at the graph of a function and attempt to graph its derivative, you are producing a graph with ordered pairs that are essentially (x, slope). If you try to do this with a great deal of numerical accuracy, it will be extremely tedious at best. Rather, you should simply estimate slopes and focus on general patterns.

Much can be learned by examining a very simple function and its derivative graphed on the same set of axes. (As you examine the graphs that follow, remember that scale has been intentionally omitted.)

Figure 11-1

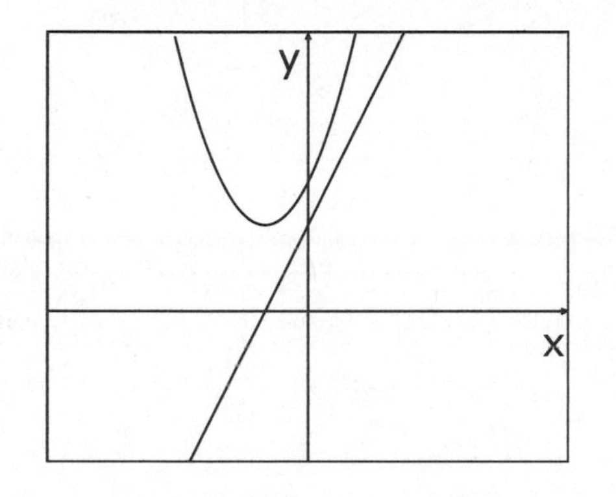

In Figure 11-1, the parabola is the function, and the line is the graph of its derivative. Symbolically this makes sense, because the derivative of a second-degree function is a first-degree function. But you are being asked to think graphically now. Imagine tangents drawn to the parabola anywhere to the left of the vertex. What kinds of slopes would those tangent lines have? Negative, of course. Examine the y values on the line directly below the parabola left of the vertex. They are all negative. To the right of the vertex, the slopes of all tangent lines are positive, and so are the y values on the line. And now here is the most important point: At the vertex, the tangent would be a horizontal line. The slope of a horizontal line is zero, and accordingly, the y value on the line is zero, an x-intercept. When producing your own derivative graphs from scratch, start by plotting the zeros of the derivative. Study Example 11-1, and before reading the explanation, try to understand how the graph of the derivative was produced.

EXAMPLE 11-1

Figure 11-2 is the graph of a piecewise function, $p(x)$, drawn with its parts labeled. The bolder graph is $p'(x)$. The five letters along the x-axis are key reference points.

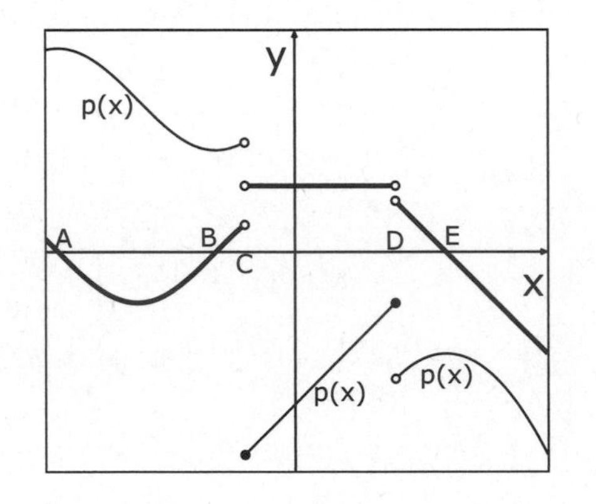

Figure 11-2

To begin, notice points A, B, and E. At those points $p'(x)$ is 0 because the tangents to $p(x)$ are horizontal. Also, at points C and D, the graph of $p(x)$ has discontinuities, so the derivative cannot exist. Thus the graph of $p'(x)$ has open circles at C and D. Just to the left of A, $p(x)$ is slightly increasing, so the graph of $p'(x)$ is positive. Between points A and B, $p(x)$ is decreasing, so any tangent drawn to the curve must have a negative slope, which makes $p'(x)$ plot below the x-axis. How far below the axis is not something to worry about right now. From B to C, $p(x)$ is sloping upward just as it did to the left of A, so $p'(x)$ again has positive values. Between C and D, $p(x)$ is linear with a constant positive slope, so $p'(x)$ is graphed as a positive constant value. From D to the end of the domain, $p(x)$ is behaving similarly to the quadratic in Figure 11-1, so its derivative graph is linear. The graph of $p'(x)$ is positive between D and E because $p(x)$ is rising in that interval.

If a function g is increasing and differentiable over an open interval (a, b), then $g'(x) > 0$ in that interval. If g is decreasing and differentiable over an open interval (c, d), then $g'(x) < 0$ in that interval.

Producing a Second-Derivative Graph

Producing a graph of a second derivative is done exactly the same way you produce the first-derivative graph. Once you have a first-derivative graph, visualize the tangent lines to that graph, estimate their slope values as zero, large positive, small positive, and so on. Then plot the values of the slopes at corresponding points of the domain.

You can actually produce your own practice problems and solutions using your graphing calculator. Plot any function on your graphing calculator, and sketch it on a piece of paper. Then draw what you believe is the derivative graph. To check, use the nDeriv(method for graphing a derivative. This method was introduced in Chapter 6. If the scale of your derivative graph is not the same as that of the derivative produced on the calculator, do not worry. As long as the x-intercepts are in similar locations and the graphs correspond in general behavior, you know you have sketched a good derivative graph.

Sketching a Function Using Its Derivative

If you have made a strong connection between a graph of a function and a graph of its derivative, the path ahead should be smooth. It is time to reverse the process and produce a rough sketch of a possible function, given information about the first derivative of the function. Again, exact function values are not a concern at this time. Nor particularly is curvature. Unless there is some information that leads you to believe the graph of the function has discontinuities, corners, or any other anomalies, your goal is just to draw a smooth differentiable function whose general behavior is revealed by its derivative.

In accordance with what you have already discovered, intervals where the derivative is positive will indicate rising portions of the function, and intervals where the derivative is negative will signal decreasing behavior on the function. The challenge will be correctly interpreting unusual behavior on the derivative graph, such as discontinuities. Before reading on, use your graphing calculator to graph $y = \ln(x^2 + 1)$. Set your window to a decimal window by pressing ZOOM followed by 4. Consider the graph that is produced to be the derivative of some unknown function *f*. Try to sketch a graph of f, starting your sketch in Quadrant III. The solution is explained in Example 11-2.

EXAMPLE 11-2

Figure 11-3 shows the graph of $f'(x)$. Assuming that $f(x)$ starts in Quadrant III, sketch a possible graph of $f(x)$.

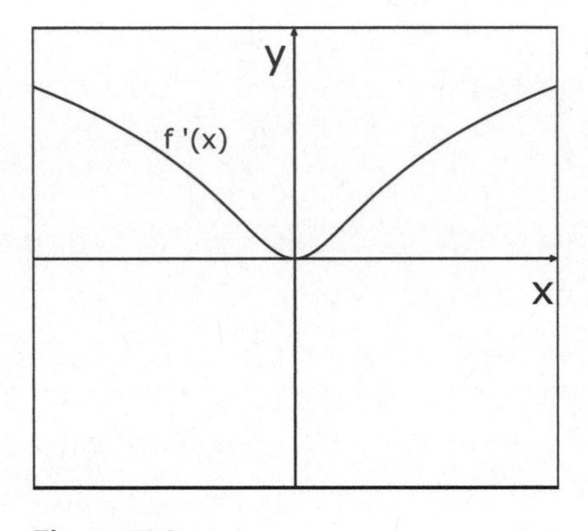

Figure 11-3

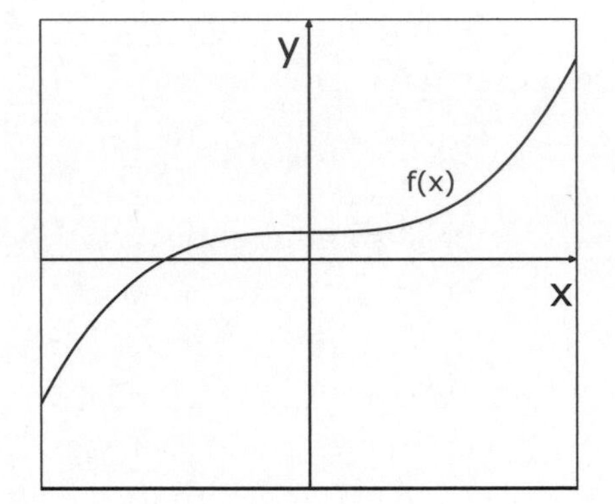

Figure 11-4

Concentrate on interpreting the y values on $f'(x)$ and what they tell you about the behavior of $f(x)$. Figure 11-3 shows that all values of $f'(x)$ are positive except at (0, 0), where $f'(x)=0$. This means $f(x)$ must be continually increasing except at $x=0$. At that point, $f(x)$ must have a tangent line with a slope of zero. Your experience may lead you to believe there should be a low or a high point there, but if a graph has a plateau at a given point, its derivative can equal zero. The non-negative y values on $f'(x)$ indicate that $f(x)$ never decreases. Neither where your attempt at this graph crosses the y axis nor the exact curvature of your sketch is currently important. If you drew a continually increasing function with a flat section on the y axis, as shown in Figure 11-4, you were successful.

Using Figure 11-5 as the graph of the derivative, try to sketch $h(x)$ as a continuous function. An explanation, along with one possible graph of $h(x)$, follows in Example 11-3.

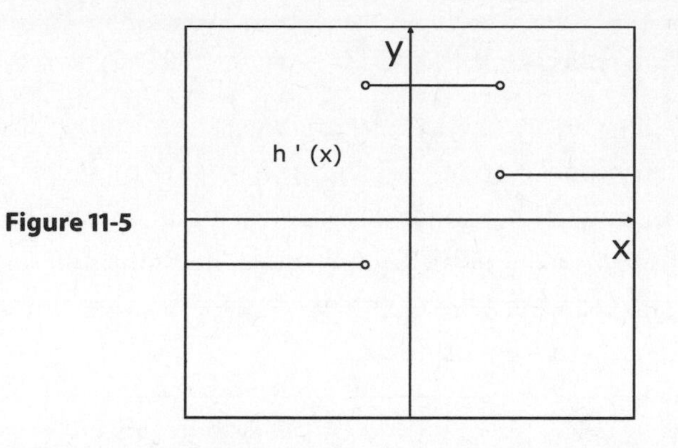

Figure 11-5

EXAMPLE 11-3

Graph a continuous function, $h(x)$, given the graph of $h'(x)$ in Figure 11-5.

Because $h'(x)$ consists of three segments, each with constant y values, the graph of $h(x)$ must have three sections, each with constant slope. The discontinuities on $h'(x)$ indicate corners or cusps—sudden changes in slope instead of smooth changes in slope. The first section of $h(x)$ must have negative slope, the second section a larger positive slope, and the third section a slightly smaller positive slope. Figure 11-6 is one possible solution.

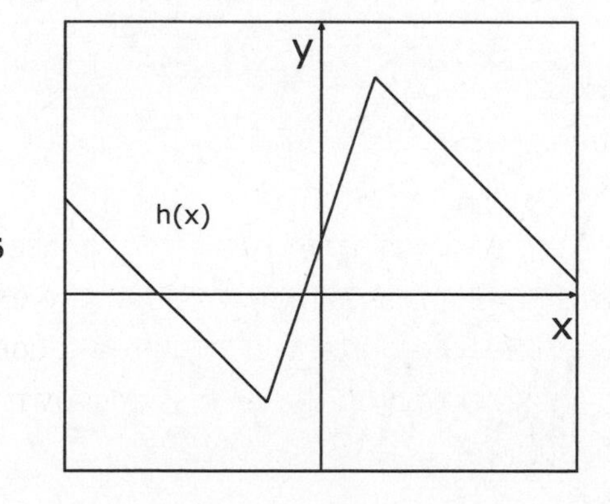

Figure 11-6

Any graph that looks like Figure 11-6 but is shifted vertically is considered a correct graph for this example.

Producing a More Detailed Graph

Up to this point, you have learned to use the first derivative to determine intervals where a graph is increasing and decreasing. With one more piece of information, you can produce the proper curvature for a graph. The information you need is the rate at which the derivative is changing. You guessed it! You need the second derivative. The second derivative determines which way a graph is curving, up or down.

> For any function g, if $g''(x) > 0$ on an open interval (a, b), then g is concave up on that interval. For any function g, if $g''(x) < 0$ on an open interval (c, d), then g is concave down on that interval.

For instance, a graph that is increasing at an increasing rate must be rising and curving up. This is because the slopes of the graph are positive and getting bigger all the time. The proper name for the curvature of a function is *concavity*. Nonlinear functions are either concave up or concave down on various intervals. In fact, all differentiable graphs that are not linear are made up of four basic shapes. Those shapes are shown in Figure 11-7. Study them carefully. They will be referenced frequently.

Figure 11-7

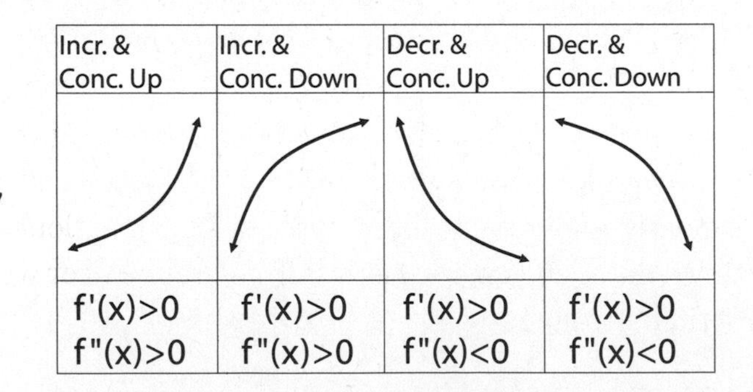

The three graphs—those of a function and its first two derivatives—are intricately intertwined. An important skill is determining which characteristics of one can be read from the others. Figure 11-7 used the graph of a function to identify values of its two derivatives on the basis of increasing and decreasing behavior and concavity. You can also use a first-derivative graph to discover information about the function and the second derivative.

Figure 11-8 is a simple portion of a first-derivative graph. Examine it and try to determine what it tells you about the function, and what it tells you about the second derivative of the function.

Figure 11-8

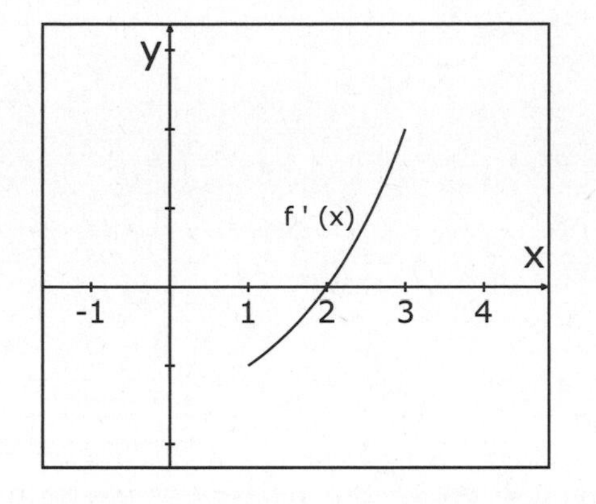

First consider just the function values of $f'(x)$. On the interval (1,2), $f'(x)$ has negative values. Because $f'(x)$ is the slope of $f(x)$, $f(x)$ must be decreasing on that interval. On the interval (2,3), $f'(x)$ is positive, so $f(x)$ must be increasing on that interval.

Now consider the slope of $f'(x)$. On the entire interval (1,3), the slope of $f'(x)$ is positive, because $f'(x)$ is increasing. But what else represents the slope of $f'(x)$? The slope of $f'(x)$ is its derivative, which is $f''(x)$. Since $f'(x)$ has positive slope, $f''(x)$ is greater than zero. When $f''(x)$ is positive, $f(x)$ is concave up. Because all three functions have been mentioned so frequently in this discussion, you may want to reread this section to be sure you clearly understand these important connections.

Constructing a detailed graph, whether from graphical or written information, should then be done in a simple logical order.

STEPS IN CONSTRUCTING A DETAILED GRAPH

1. Plot any known ordered pairs on the graph.
2. Use the signs of the first derivative to sketch increasing and decreasing intervals on the graph.

3. Use the signs of the second derivative to sketch in the proper intervals of concavity.

EXAMPLE 11-4

Produce a detailed sketch of a differentiable function from the following information.

$f(0) = 0$ and $f(1) = -3$

$f'(x) > 0$ for $x < -2$ and $x > 4$ and $f'(x) < 0$ for $-2 < x < 4$

$f''(x) < 0$ for $x < 1$ and $f''(x) > 0$ for $x > 1$

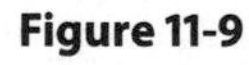

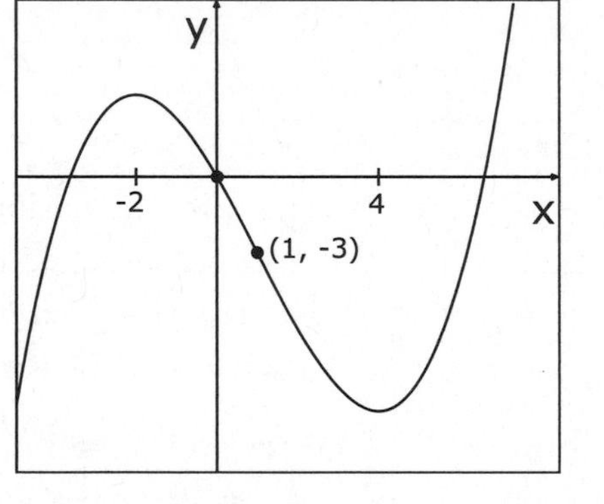

Figure 11-9 satisfies the various conditions. $f(x)$ contains the proper points. The graph is increasing for $x < -2$ and $x > 4$, because its derivative is negative. $f(x)$ is also increasing for $-2 < x < 4$, where its derivative is positive. To the left of $x = 1$, the graph of $f(x)$ is concave down, as required by $f''(x) < 0$ for $x < 1$. To the right of $x = 1$, the graph is concave up.

The final example uses a sketch of the derivative of a function to produce a sketch of the function itself. Specific points on the function are not given, because the focal point at this time is general behavior and accuracy of shape.

EXAMPLE 11-5

Use the sketch of $h'(x)$ over the domain $[A, D]$ in Figure 11-10 to write a description of the graphical behavior of $h(x)$. Address increasing intervals, decreasing intervals, and intervals of concavity. Then produce a correctly shaped sketch of $h(x)$.

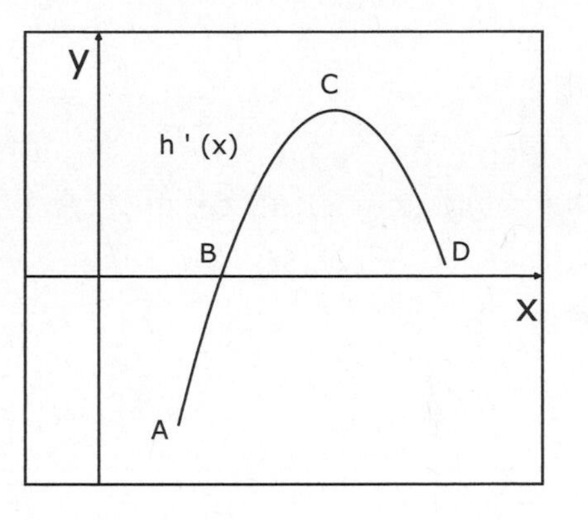

Figure 11-10

From A to B, $h'(x)$ is negative, so $h(x)$ must be decreasing. But $h'(x)$ is increasing from A to B, so $h''(x)$ must be positive, making $h(x)$ concave up. From B to C, $h'(x)$ is positive and increasing, so $h(x)$ is increasing and still concave up. From C to D, $h'(x)$ is positive, so $h(x)$ is still increasing. But from C to D, $h'(x)$ is decreasing, so $h''(x)$ is negative, which makes $h(x)$ concave down. See Figure 11-11.

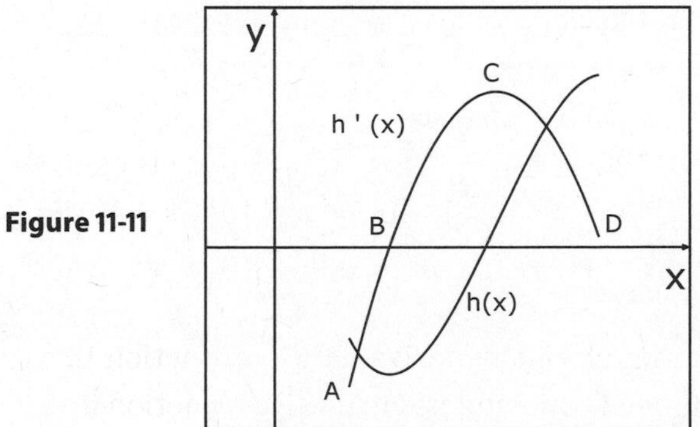

Figure 11-11

Skill Check

For some additional practice, try the problems that follow. Remember to carefully interpret the relationships between the function and its first and second derivatives. Your primary goal is to use the derivatives to gain insight into the behavior of the function, but remember that the first derivative can also give you information about the second derivative.

1. By hand, sketch $y = x^{\left(\frac{2}{3}\right)}$ and its derivative on the same set of axes.
2. If $g'(x)$ is negative and increasing on an interval, describe the behavior of $g(x)$, addressing its increasing or decreasing and its concavity.
3. Figure 11-12 shows a graph of $h'(x)$. On the same interval, sketch $h''(x)$.

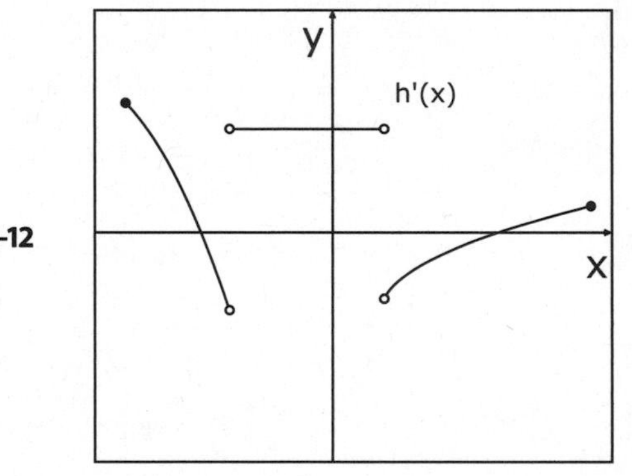

Figure 11-12

4. Figure 11-13 shows a graph of $p'(x)$. On the same interval, sketch a graph of $p(x)$ as accurately as you can.

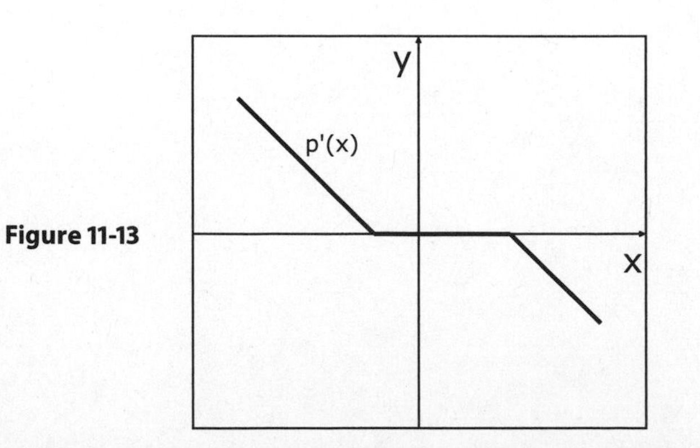

Figure 11-13

5. Given the following information, sketch a continuous graph of $r(x)$ as accurately as you can on the domain $[-2,3]$.

   $r(-2)=0$, $r(1)=3$, and $r(3)=1$

   $r'(x)>0$ on the interval $[-2,1]$. $r'(x)<0$ on the interval $[1,3]$. $r'(1)$ is undefined.

   $r''(x)>0$ on the interval $[-2,1]$. $r''(x)<0$ on the interval $[1,3]$.

CHAPTER 12

Applications of Derivatives

Will states eventually use electronic monitoring as a way to determine when a driver has operated his vehicle at excessive speed between tollbooths and therefore issue a ticket? When is an epidemic growing fastest and when is it finally going to begin to decline? These questions and countless more are answered using calculus principles. With a few more tools that you will learn in this chapter, you will understand many of the multitude of applications that can be explored and analyzed using calculus.

Local Maxima and Minima

A brief foray into maximum and minimum values of a function can lay the groundwork for exploring almost any optimization application. One of the key tools used in optimization problems is determining local maximum or minimum values for a function. Simply stated, a local maximum (or local minimum) value of a function is the highest (or lowest) point on a localized domain. Given a function $f(x)$ and a point $(a, f(a))$, then $f(a)$ is a local maximum if there exists some open interval containing $x = a$ such that all function values in the interval are less than or equal to $f(a)$. Likewise, a local minimum must be less than or equal to all function values in the interval.

A critical value of a function $g(x)$ is any x value where $g'(x)=0$ or where $g'(x)$ is undefined. These are the locations where local extrema *may* exist.

To search for a local maximum or minimum on a function, you first find the critical values. These are the x values you test to determine whether any of them are local maxima or minima. In order for a local minimum to occur at a critical value on a function, the function must be continuous, and the sign of the derivative must change from negative to positive at the critical point. As you should know from the previous chapter, the function would be decreasing to the left of the critical point and increasing to the right, thereby creating a local minimum value. Again, assuming continuity, a local maximum occurs when the derivative changes from positive to negative. A good organizational tool is a sign chart on a number line. Because the Intermediate Value Theorem guarantees that the derivative cannot change signs between critical points without creating another critical point, testing the sign of the derivative in between critical points is very informative.

EXAMPLE 12-1

Find the x-coordinate of each local extreme value on the graph of $f(x) = x^3 - x^2 - 5x + 1$.

$f'(x) = 3x^2 - 2x - 5$ Take the derivative.

$3x^2 - 2x - 5 = 0$	Find where the derivative equals zero.
$(3x-5)(x+1) = 0$	Solve by factoring.
$x = -1$ or $x = \frac{5}{3}$	

Pick any x value less than –1, any -value between –1 and 5/3, and any x value larger than 5/3. Test each number in $f'(x)$ and determine the sign. The exact value is not needed. Record the signs between the critical values, as shown in Figure 12-1. For example, in the factored form, $f'(-2) = (-6-5)(-2+1) > 0$.

Figure 12-1

Because $f'(x) > 0$ for $x < -1$, $f(x)$ is increasing on that interval. Because $f'(x) < 0$ for $-1 < x < \frac{5}{3}$, $f(x)$ is decreasing on that interval. $f(x)$ goes from increasing to decreasing at $x = -1$. This indicates that a local maximum exists on $f(x)$ at $x = -1$. Similar reasoning is used to determine that $f(x)$ has a local minimum at $x = \frac{5}{3}$. To confirm this result, graph the function on your calculator.

Be especially careful as you examine function behavior at critical points when a function or its derivative has discontinuities. The function's derivative will be undefined, and you will have to check signs on each side of those critical points.

Not all critical points of a function are locations of relative extrema. In order for a function to have a local maximum or minimum at a point on the interior of its domain, the derivative of the function must change signs at that point.

EXAMPLE 12-2

Find the x-coordinate of each local extreme value on the graph of $f(x)=2+\sqrt[3]{x}$.

Think of $f(x)$ as $f(x)=2+x^{\left(\frac{1}{3}\right)}$.

$f'(x)=\frac{1}{3}x^{\left(\frac{-2}{3}\right)}=0$ Using the power rule, set the first derivative equal to zero.

$f'(x)=\frac{1}{3\sqrt[3]{x^2}}=0$ Rewrite the derivative in a more useful form.

The only critical value is $x=0$ because $f'(x)$ is undefined at that point.

$f'(-1)=\frac{1}{3}$ and $f'(1)=\frac{1}{3}$, so the derivative does not change sign. $f(x)=2+\sqrt[3]{x}$ has no local extrema. If you graph $f(x)=2+\sqrt[3]{x}$ on your calculator, you will see that the function actually has a vertical tangent at the point $(0,2)$. Thus the function is continuous, but the limit of the slope of the tangent as x approaches zero is infinite.

Second-Derivative Test

There is one other way to test for a local maximum or minimum at a point. It is commonly called the second-derivative test for extrema. If the second derivative is easy enough to find, this test can be very convenient. At a given point $x=c$, if $f'(c)=0$ and $f''(x)>0$, then the function has a local minimum. If $f'(c)=0$ and $f''(x)<0$, then the function has a local maximum. Just think about a parabola that opens upward, such that the vertex is a local minimum. At the vertex there is a horizontal tangent, and the function is concave up. The first derivative is zero, and the second derivative is positive. Note how this works on the function we first encountered in Example 12-1 because the second derivative is easy to find.

EXAMPLE 12-3

Find the x-coordinate of each local extreme value on the graph of $f(x)=x^3-x^2-5x+1$.

$f'(x)=3x^2-2x-5$

$3x^2-2x-5=0$ Find where the derivative equals zero.

$x=-1$ or $x=\frac{5}{3}$

$f''(x)=6x-2$ Find the second derivative.

$f''(-1)=6(-1)-2<0$ Test the second derivative at each point where $f'(x)=0$.

$f''\left(\frac{5}{3}\right)=6\left(\frac{5}{3}\right)-2>0$

Because $f'(-1)=0$ and $f''(-1)<0$, there is a local maximum at $x=-1$.

Because $f'\left(\frac{5}{3}\right)=0$ and $f''\left(\frac{5}{3}\right)>0$, there is a local minimum at $x=\frac{5}{3}$.

Absolute Extrema

Another often-used tool in applications is an absolute extreme. An absolute maximum (or absolute minimum) is, as you would imagine, the greatest (or least) function value anywhere on a closed interval or on the entire function, depending on the situation. On any closed interval, a function has an absolute highest point and an absolute lowest point. This is not guaranteed if the domain of a function is open. In those situations, knowing what the basic graphs look like can be very helpful in determining whether an absolute highest or lowest point exists. For instance, a square root function always has an absolute minimum value, but if it is reflected over the x-axis, it then has an absolute maximum value.

On closed intervals, the process of finding absolute extrema is just a bit longer than finding local extrema. After finding the local extrema, you must compare their function values to the function values of the endpoints. On a closed interval, the local maxima, local minima, and endpoints are the only candidates for the absolute extrema. It is also important to note that endpoints of a closed interval can be local extremes. Additionally, all absolute extremes are also local extremes. Example 12-4 takes you through the process.

EXAMPLE 12-4

Find the local and absolute extremes of the function

$h(x) = \frac{1}{3}x^3 - x^2 - 3x - 4$ on the interval [−2,6].

$h'(x) = x^2 - 2x - 3 = 0$ Find the local extreme values.

$(x-3)(x+1) = 0$, so $x = 3$ or $x = -1$

$h''(x) = 2x - 2$

$h''(-1) = 2(-1) - 2 < 0$ By the second-derivative test, $h(x)$ has a local maximum at $x = -1$.

$h''(3) = 2(3) - 2 > 0$ By the second-derivative test, $h(x)$ has a local minimum at $x = 3$.

Compare function values at local extrema to function values at the endpoints.

$h(-2) = -4\frac{2}{3}$, $h(-1) = -2\frac{1}{3}$, $h(3) = -13$ and $h(6) = 14$.

Local minimum values exist at, $h(-2)$ and $h(3)$. Local maximum values exist at $h(-1)$ and $h(6)$.

The absolute minimum value is −13 at $x = 3$. The absolute maximum value is 14 at $x = 6$.

Optimization

All rational business investors want to maximize their profits. One way to achieve that is to reduce costs of production. Within production, minimizing the cost of packaging may be one way to increase profits. Putting machinery or even entire production facilities in just the right locations may also reduce production or shipping costs. Granted, some of these problems may fall into the area of discrete math, but at the heart of it all are calculus and the application of maximization and minimization skills.

The following list will help you stay organized as you tackle and master optimization problems.

1. Read and understand what is being sought.
2. Make a carefully labeled diagram, if applicable.
3. Write down formulas and given information.
4. Write a function for what is to be optimized.
5. Set the derivative of the function equal to zero and solve.

Most optimization situations have only one reasonable solution, but if you do get two or more solutions, use a signed-number line or the second-derivative test to determine maximums and minimums. Also, always make sure your answer makes sense and is in the reasonable domain of the problem.

EXAMPLE 12-5

A goat farmer wants to build a rectangular pen adjacent to a barn. The pen must be 2,700 square feet in area. Because of the need to prepare the land, the side parallel to the barn will cost one and a half times as much per foot to build as the two sides perpendicular to the barn (see Figure 12-2). What dimensions will produce a 2700-square-foot pen for the lowest cost?

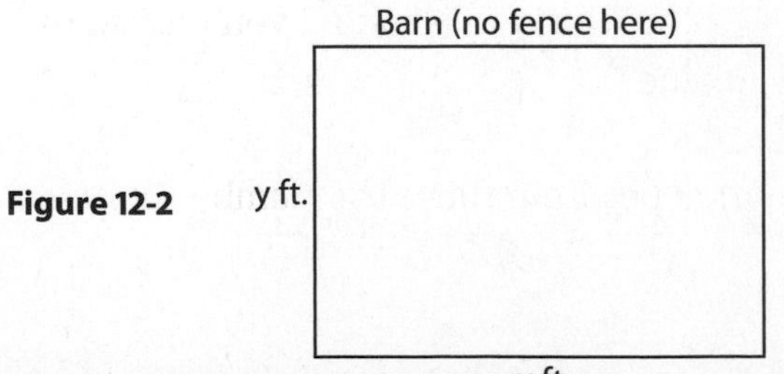

Figure 12-2

The equation for the area is $xy = 2700$. But you want to minimize the cost of the fence.

The cost of the side labeled x is one and a half times the cost of the side labeled y.

The unit cost of the fence will not matter, as long as the proportions are correct. The cost function is $C = 1.5x + 2y$.

Make the cost function one variable by substituting from $xy = 2700$, or $y = \frac{2700}{x}$.

$$C = 1.5x + \frac{5400}{x} = 1.5x + 5400x^{-1}$$

$$C'(x) = 1.5 - 5400x^{-2}$$

$$1.5 - 5400x^{-2} = 0$$

Solving the equation yields $x = 60$ feet and $y = \frac{2700}{60} = 45$ feet. The side parallel to the barn should be 60 feet, and each of the sides perpendicular to the barn should be 45 feet.

The next example requires no diagram. It is a business application problem, but the process remains the same.

EXAMPLE 12-6

The number of items you sell is a function of the price you charge. Say that if x represents the price you charge per item, then the number you sell is modeled by $N = 2000 - 3 \cdot 2^x$. What price should you charge to make the greatest amount of income?

The revenue you make is the price per item times the number of items you sell, or $x \cdot N$.

Let revenue be $R(x) = x(2000 - 3 \cdot 2^x)$.

Find the derivative by the product rule and set it equal to zero.

$R'(x) = x(-3 \cdot \ln(2) \cdot 2^x) + (2000 - 3 \cdot 2^x) \cdot 1 = 0$

This is an example of an equation that you need to solve using your graphing calculator.

Graph $R'(x)$ on your calculator, and find the x-intercept using the ZERO finder.

The graph of the derivative goes from positive to negative at about $x = 6.86$. Therefore, the revenue function will be a maximum when items are sold for \$6.86 each.

Optimization problems are immensely varied, and a whole book could be dedicated to these types of problems alone. The goal of this section is to introduce the process of solving these problems and to give you a little practice. If you have a particular field of interest, do a search for optimization problems in that field. You are likely to find plenty of sources of more practice to broaden your experience.

Inflection Points

Another point of interest in applications of the derivative is the time a quantity is changing the fastest. As the seasons of the year come and go, the hours of daylight are constantly changing, but not in a steady manner. There are times of year when the rate of change of daylight is a maximum and other times when it is a minimum.

As you know by now, the rate of change of a function, call it $g(x)$, is its derivative $g'(x)$. When you are looking for where $g'(x)$ is the greatest, you are treating $g'(x)$ as just another function and trying to maximize (or in some cases minimize) that function. Extreme points on any function happen where the derivative of the function changes from positive to negative or from negative to positive. In this case, $g'(x)$ is the function, and it is the second derivative, $g''(x)$, that is changing signs. Now think about what this implies for the original function, $g(x)$. The sign of the second derivative determines concavity. If the second derivative changes signs, then the function $g(x)$ is changing concavity. The point where this happens is called an inflection point. Figure 12-3 shows an inflection point on a function, along with the first and second derivatives of the function.

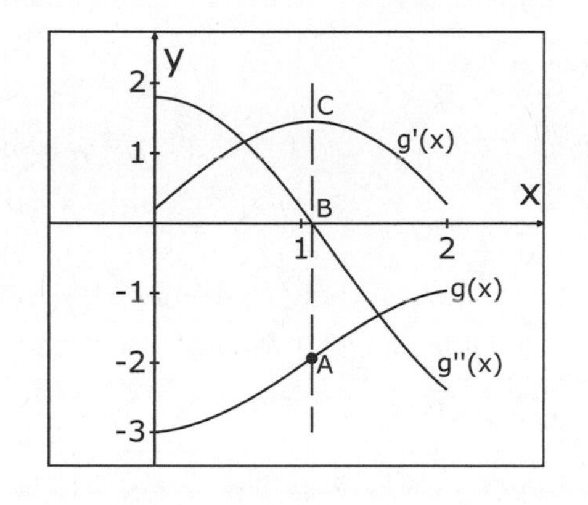

Figure 12-3

Note that at point A, $g(x)$ has an inflection point. At the same x value, point B shows $g''(x)$ changing from positive to negative. That corresponds with $g(x)$ changing from concave up to concave down, and with $g'(x)$ going from increasing to decreasing at point C, its local maximum.

During a chemical reaction, when a solution is changing from a base to an acid or from an acid to a base, the rate of reaction is greatest at the instant that the pH changes sign. On a graph of the changing pH, this is the inflection point.

EXAMPLE 12-7

Find the x-coordinate of the inflection point on $y = 3e^{(-x^2)}$.

Find the point where the second derivative changes sign.

$y' = -6x \cdot e^{(-x^2)}$ Use the derivative rule $\frac{d}{dx}(e^u) = e^u \frac{du}{dx}$.

$y'' = -6(x \cdot -2xe^{(-x^2)} + e^{(-x^2)} \cdot 1)$ Use the product rule on y'.

$-6e^{(-x^2)}(-2x^2 + 1) = 0$ Factor out $e^{(-x^2)}$ and set y'' equal to zero.

$-2x^2 + 1 = 0 \qquad x = \pm\sqrt{\frac{1}{2}}$ $e^{(-x^2)}$ is never zero, so solutions come from $-2x^2 + 1 = 0$.

Use a graphical method to determine that the second derivative changes sign at those x-coordinates. The graph of $y = -2x^2 + 1$ is a parabola, which opens downward. It crosses the x-axis and thus changes sign at $x = \pm\sqrt{\frac{1}{2}}$, guaranteeing inflection points.

A word of caution is necessary here. Not every change in the sign of a second derivative produces an inflection point on a function. The function must be continuous at the point where the second derivative changes sign. A simple example is the graph of $y = \frac{1}{x}$. It changes concavity at $x = 0$ but has a vertical asymptote there, not an inflection point.

One other scenario that must be mentioned is when the second derivative is undefined at a point and changes sign there. As long as the function has a tangent line at the point in question, there will be an inflection point. For example, $y = \sqrt[3]{x}$ has an undefined second derivative at $x = 0$ but has an inflection point at (0, 0). Ultimately, careful analysis of what is happening with the function and with the second derivative will lead to a proper conclusion.

The Mean Value Theorem

Have you ever driven on a toll highway where you pick up a card as you get on the road and then present the card when you exit? The system tracks where you entered and where you left to determine how much toll you pay.

But did you also notice that it records the time at each checkpoint? Consider what this means. If, for example, you travel 280 miles in 4 hours, your average speed is 70 miles per hour. But what if the speed limit on that particular highway is 60 miles per hour? The state could use calculus to prove you were speeding!

The common-sense explanation is that there is no way you can average 70 miles per hour if your speed is always below 70 miles per hour. At some point you had to be going at least 70 miles per hour! This essentially is what the Mean Value Theorem says in more formal terms.

The Mean Value Theorem: If a function $f(x)$ is continuous on the closed interval $[a,b]$ and is differentiable on the open interval (a,b), then there exists at least one value c in (a,b) such that $f'(c)=\frac{f(b)-f(a)}{b-a}$.

Thinking back to earlier chapters will help you understand this from another perspective. The left-hand side of the equation, $f'(c)$, is a slope of a line tangent to the function, or the instantaneous rate of change of f. The right-hand side of the equation is the slope of the segment between the points $[a,f(a)]$ and $[b,f(b)]$, which is the average rate of change of the function on the interval. The theorem states that the instantaneous rate of change must equal the average rate of change at least once in the interval. If the function is speed, then the instantaneous speed must equal the average speed at least once during a trip.

EXAMPLE 12-8

Find all values of c in the interval $[e^{(-2)},e]$ that satisfy the Mean Value Theorem for $g(x)=\ln(x)+2$. You will need to evaluate on your calculator.

$g[e^{(-2)}]=\ln[e^{(-2)}]+2=-2+2=0 \qquad g(e)=\ln(e)+2=1+2=3$

$$g'(x)=\frac{1}{x}$$

$$\frac{1}{c}=\frac{3-0}{e-e^{-2}}$$

Set the derivative evaluated at c equal to the slope between $(e^{(-2)},0)$ and $(e,3)$.

$$c=\frac{e-e^{(-2)}}{3}\approx 0.861$$

Write the reciprocal of both sides of the equation.

Figure 12-4 shows this is the point where the slope of the tangent line equals the slope of the secant line.

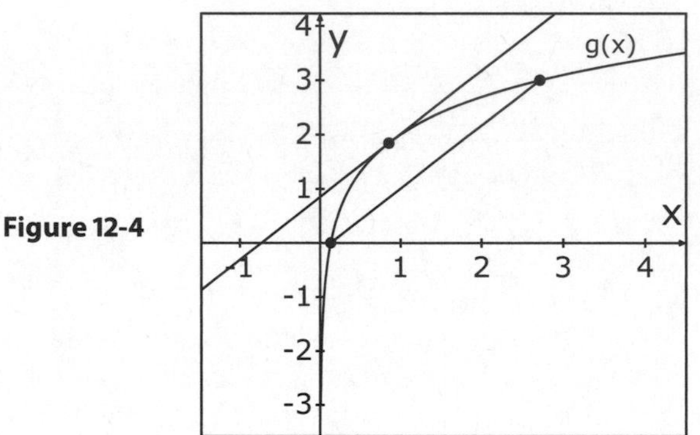

Figure 12-4

Linear Motion

Chapter 10 touched on linear motion, but it is worth revisiting this topic that is so important to physics. For a first course in calculus, the motion is always along a line. The next course examines motion in the plane. Recall that the first derivative of the position function is the velocity function. The rate of change of velocity, which is known as acceleration, is the first derivative of velocity and the second derivative of position.

Those who work with motion take great interest in the extrema of these functions. When is velocity at its greatest? When does the particle change directions? You can now view these questions in the context of maximums, minimums, and inflection points to give you a better understanding of the motion and behavior of the moving object.

EXAMPLE 12-9

A particle moves along a horizontal path such that its directed distance from a fixed point is modeled by $s(t)=t^3-7t^2+8t+3$ for $t\geq 0$. Time is measured in seconds and distance in feet. Define left of the fixed point as negative position and right of the fixed point as positive position. Describe the motion of the particle, and determine when its velocity is the most negative.

$v(t)=s'(t)=3t^2-14t+8$ and $a(t)=v'(t)=6t-14$

$v(t)=(3t-2)(t-4)=0$ — Find where the velocity is zero, where it is positive, and where it is negative.

$v(t)>0$ for $0\leq t<\frac{2}{3}$ and $t>4$ — Use a signed-number line or the graph of velocity.

$v<0$ for $\frac{2}{3}<t<4$

$a(t)=6t-14=0$ when $t=\frac{7}{3}$.

$a(t)<0$ for $t<\frac{7}{3}$ and $a(t)>0$ for $t>\frac{7}{3}$.

The velocity results tell you the particle starts out moving to the right, because $v(t)>0$ when $t=0$. At $t=\frac{2}{3}$ seconds, the velocity changes sign, so the particle turns and goes to the left until $t=4$ seconds. At this point it turns and goes to the right, because velocity changed back to positive.

Acceleration is the derivative of velocity, so acceleration changing from negative to positive tells you the velocity is a minimum at $t=\frac{7}{3}$ seconds.

The position graph and the velocity graph are shown in Figure 12-5. When the position graph is increasing, velocity is positive, and when the position graph is decreasing, velocity is negative. Note also that the velocity is at a minimum at the same value of t at which the position graph has an inflection point. Concavity of the position graph corresponds to the sign of acceleration.

Figure 12-5

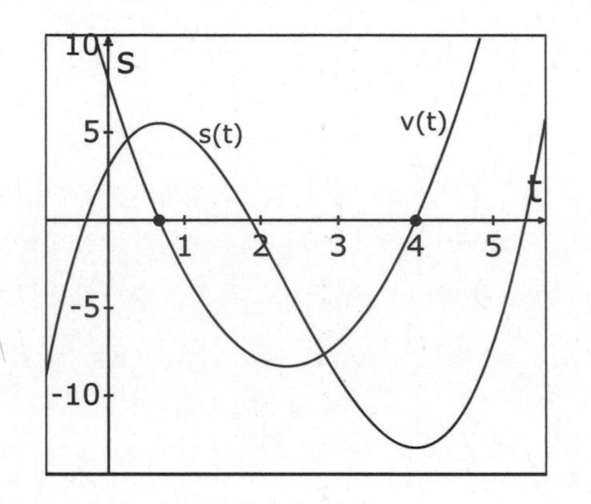

Related Rates

Another important application of derivatives is related rates. This is essentially the study of how multiple quantities changing at an instant in time are connected. A simple but practical application is a police officer driving toward you as you drive toward him. If his speed is 35 miles per hour, and your speed is 32 miles per hour, the radar gun in his car will report how fast the distance between cars is diminishing. In this case, the radar will read 67 miles per hour. If it reads too high, the officer can subtract his speed and determine you are speeding!

Like optimization problems, problems that involve related rates are best solved with a systematic approach. The challenge is that solving them entails drawing on a wide range of prerequisite knowledge to determine relationships between quantities and to make the proper connections between the variables in the problem. It also requires a form of implicit differentiation, because the rates of change of all quantities are examined with respect to time.

Here is a helpful approach to follow:

1. Read and understand the problem.
2. Make a carefully labeled diagram.
3. Write down and label constants, variables, rates, and what is being sought.
4. Write a function that relates the variables.
5. Differentiate all terms with respect to time.
6. Substitute known quantities and solve.

Are you ready to try a few? The following examples will help you get started. As you begin, always remember two key points. First, every rate is going to be a derivative with respect to time, *t*. Second, never substitute a value for a changing quantity before differentiating.

EXAMPLE 12-10

Air is being blown into a spherical balloon at a rate of 4 cubic inches per minute. How fast is the radius of the balloon increasing the instant that the radius is 6 inches?

The change in radius is being examined when $r = 6$ in.

The air being blown in is a rate of change of volume, so $\frac{dV}{dt} = 4\ \frac{\text{in}^3}{\text{min}}$.

The volume and radius of a sphere are related by $V = \frac{4}{3}\pi r^3$.

$\frac{dV}{dt} = \frac{4}{3}\pi \cdot 3r^2 \frac{dr}{dt}$ Differentiate the volume formula with respect to time.

$4 = \frac{4}{3}\pi \cdot 3(6)^2 \frac{dr}{dt}$ Substitute the known quantities after differentiating.

$\frac{dr}{dt} = \frac{1}{36\pi}\ \text{in}/\text{min}$ The units are inches per minute because the radius is a linear measure.

Example 12-11 provides one more look at a related-rate problem with a bit more detail.

EXAMPLE 12-11

A zero-depth pool is angled downward at 18° (see Figure 12-6). You are walking steadily toward the deeper water at a rate of 2 feet per second. At the instant you are 12 feet from the edge of the water, how fast is the water level rising on you?

Figure 12-6

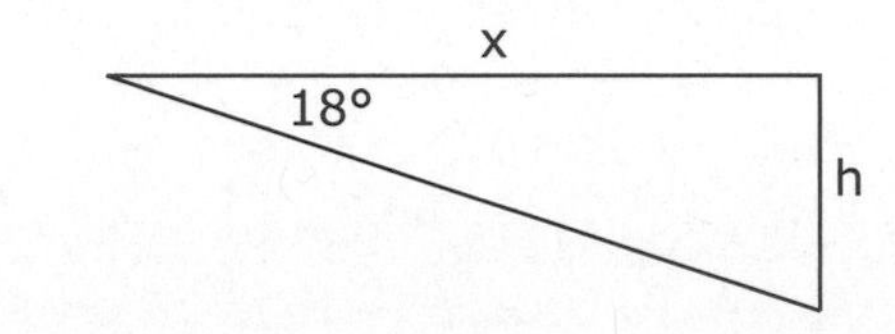

Let x be the distance you have walked into the pool, and let h be the depth of the pool.

At the instant you are examining, $x = 12\text{ ft}$ and $\frac{dx}{dt} = 2\ {}^{\text{ft}}\!/_{\text{sec}}$. You could calculate h, but it is not needed at this point. You are seeking $\frac{dh}{dt}$, the rate of change of depth with respect to time.

The variables are related by $\tan(18^\circ) = \frac{h}{x}$, or $x \cdot \tan(18^\circ) = h$.

$$\frac{dx}{dt}\tan(18^\circ) = \frac{dh}{dt}$$ Differentiate with respect to time.

$$\frac{dh}{dt} = 2\tan(18^\circ) \approx 0.65\ {}^{\text{ft}}\!/_{\text{sec}}$$

It was okay to insert the $\tan(18^\circ)$ value into the equation before differentiating, because the angle of the pool never changed. By contrast, the 12, which was the value of x, could not be substituted before differentiating. Interestingly, the 12 never mattered, which means the water depth was changing at a steady rate the whole time.

Skill Check

There are a vast array of problems from which to choose for practice in this chapter. Because you are probably either reviewing calculus after a number of years away or seeing it for the first time, this set of problems will help you practice the basics without delving too deeply into complicated problems. If you are feeling confident and want more of a challenge, consult almost any standard calculus textbook for an abundance of rigorous problems on all of these topics.

1. Find the local and absolute extreme values on $k(x)=(x-2)^2+3$ on the interval $[0,5]$.
2. Find the x-coordinate of each inflection point of the function $h(x)=\ln(x^2+1)$.
3. Use the graph of $w'(x)$ in Figure 12-7 to determine the x-coordinates where $w(x)$ has local extremes and an inflection point.

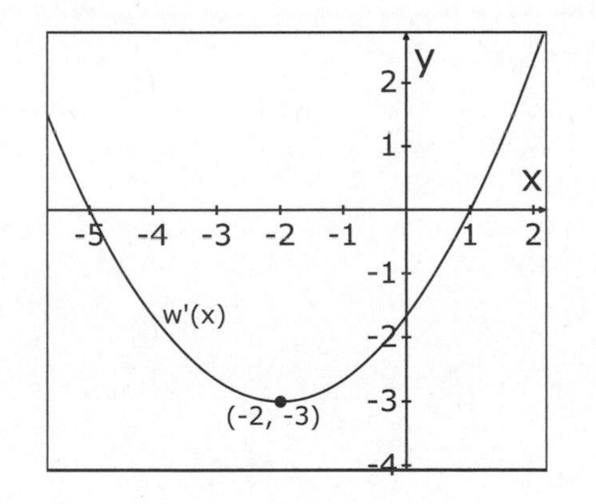

Figure 12-7

4. A sheet of cardboard has dimensions 6 inches by 6 inches. Squares of equal size will be cut from each corner of the box, and the remaining tabs will be folded up to make an open-top box. What size squares should be cut from the corners to create the box with the greatest volume?
5. Find the value of c that satisfies the Mean Value Theorem for $f(x)=\dfrac{2^x}{x}$ on the interval $[1,4]$. After setting up the equation, solve it with your calculator.

6. A particle moves along the x-axis such that its position as a function of time is $x(t)=\frac{1}{3}t^3-9t+1$ for $t\geq 0$. Find the particle's position and acceleration the first time its velocity is zero.
7. Two trains are approaching the same station. One train is traveling on a southbound track at 40 miles per hour and is 30 miles from the station. A second train is traveling on an eastbound track at 60 miles per hour and is 40 miles from the station. At that instant, how fast is the straight-line distance between the trains changing?

CHAPTER 13

Area by Numerical Methods

With differential calculus under your belt, it is time to begin the second major division of a typical first-year calculus course: integral calculus. This chapter will open the door to the realm of integrals through the numerical exploration of areas between graphs and the *x*-axis. Areas under graphs can have real applied meaning. This chapter lays the foundation for what lies ahead, but it also ties new ideas to one of the first concepts you studied, limits.

Area under a Graph

One of the first algebra application formulas most people learn is the following simple relationship: rate times time equals distance. If a jet travels a constant 500 miles per hour and flies for 2 hours, then barring the effects of wind, the jet will travel 2×500 or 1,000 miles. But if the jet frequently changes speeds, calculating the distance covered may not be so easy.

Or consider a freely falling object experiencing a negligible amount of wind resistance. If you drop a stone off a high bridge and time how long it takes to hit the ground, you can determine the height of the bridge by using some basic calculus. Thanks to the laws of physics, the ever-changing velocity of the stone provides enough information for you to find the total distance the stone has fallen.

These simple scenarios illustrate the relevance and significance of being able to find area under a graph. As with so many other ideas in calculus, the knowledge pursued is for the purpose of understanding and interpreting the surrounding world.

As a first look at integral calculus, study Figure 13-1. It is a graph of the speed of a person walking at a steady rate of 3 miles per hour for 2 hours. Without the graph, you could easily conclude that the person walked 3×2 or 6 miles. But it is the geometric significance that matters here. The graph forms a rectangle with the x-axis, the dimensions of which are 3 units high and 2 units wide. The area between the velocity graph and the x-axis represents the walker's displacement.

Figure 13-1

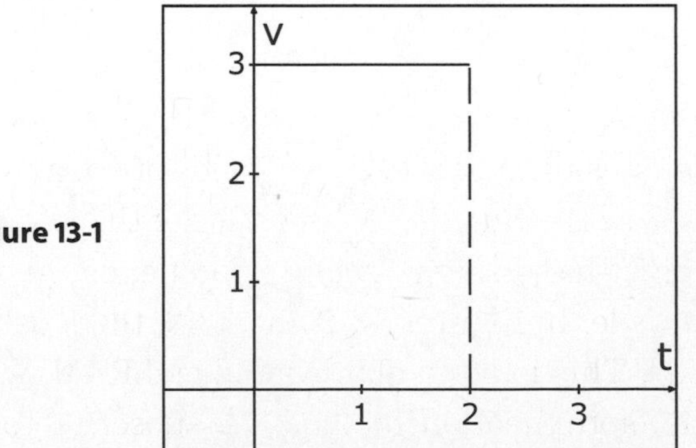

Riemann Sums

The great German mathematician Bernhard Riemann took the idea of "area under a graph" and extended it to deal with graphs of nonlinear functions. His approach was to consider the rate of change for very small intervals, treat the function as nearly constant on each interval, multiply rate times time, and sum up the resulting displacements. Essentially, he was adding up the areas of many narrow rectangles.

Look at Figures 13-2 and 13-3. Each estimates the area between the curve $f(x) = \sqrt{x-1}+3$ and the x-axis on the interval $[1,5]$. Figure 13-2 uses just two inscribed rectangles, whereas Figure 13-3 uses four inscribed rectangles. Which do you think provides the better estimate of the area under the function? You're right! As you increase the number of rectangles, the approximation of the area under the curve improves.

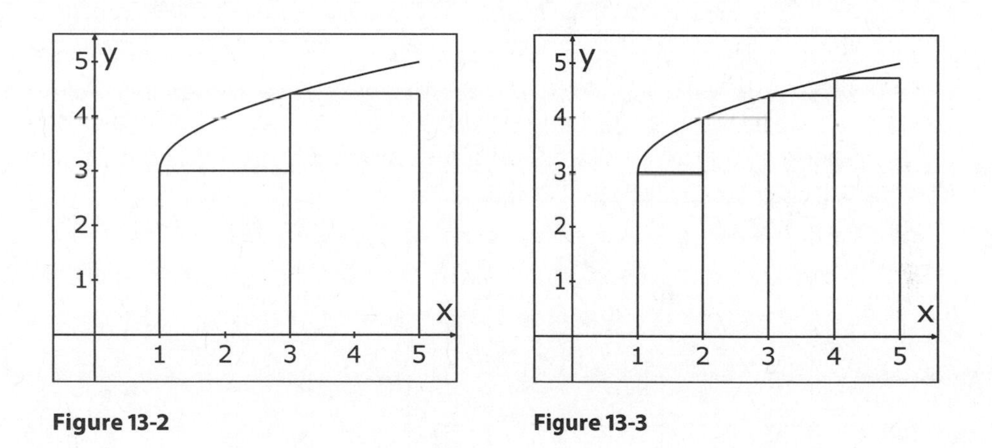

Figure 13-2 **Figure 13-3**

Riemann worked with rectangles that could be any width, but modern-day introductory calculus courses primarily use rectangles of equal width whose heights are determined by the function value at either the left-hand side of each rectangle, the right-hand side of each rectangle, or the midpoint of the base of the rectangle. In Figures 13-2 and 13-3, the left-hand side of each rectangle is used. This is often abbreviated as LRAM, which stands for "left rectangular approximation method." A subscript to the last letter is used to indicate how many rectangles should be used in the approximation. For instance, $RRAM_8$ means that you should draw eight rectangles whose heights are function values at the right-hand end of each rectangle. A little practice with each method will be useful, but the number of rectangles will be kept low to avoid too much numerical tedium.

LRAM

Returning to Figure 13-3, note that the width of each rectangle—call it Δx because it is a change in x—is 1 unit. The heights of the four rectangles are determined by $f(1)$, $f(2)$, $f(3)$, and $f(4)$. Using a calculator to help find values reveals that the sum of the areas is $\text{LRAM}_4 = 1 \cdot [f(1) + f(2) + f(3) + f(4)] \approx 16.146$. Examining the drawing makes it clear that this result is an underestimate.

RRAM

To begin generalizing the idea toward the conceptual goal, see Figure 13-4, where two more rectangles have been added and oriented on the basis of their right-hand edges. This process will take a bit more thought, and it will calculate RRAM_6. (RRAM, of course, stands for "right rectangular approximation method.")

Figure 13-4

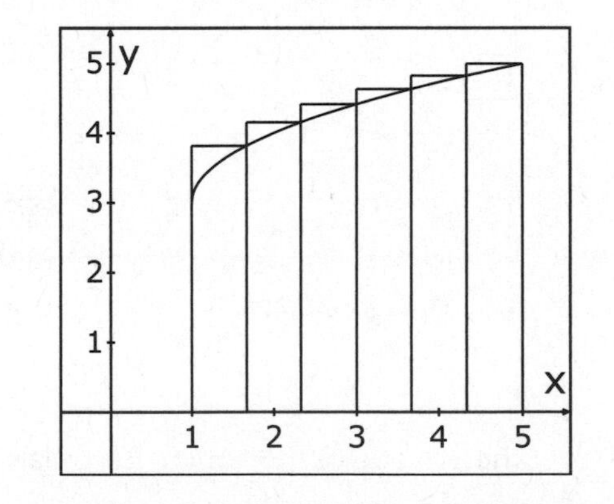

In any situation, to calculate Δx, divide the width of the interval by the number of rectangles. In this case, $\Delta x = \frac{5-1}{6} = \frac{2}{3}$. Because the heights are based on the right-hand sides of the intervals, $f(1)$ will not be used. The sum of the areas of the six triangles are $\text{RRAM}_6 = f\left(\frac{5}{3}\right) \cdot \frac{2}{3} + f\left(\frac{7}{3}\right) \cdot \frac{2}{3} + f\left(\frac{9}{3}\right) \cdot \frac{2}{3} + f\left(\frac{11}{3}\right) \cdot \frac{2}{3} + f\left(\frac{13}{3}\right) \cdot \frac{2}{3} + f\left(\frac{15}{3}\right) \cdot \frac{2}{3}$. It is repetitive to write the Δx each time, so it is factored out to produce

$$\text{RRAM}_6 = \left[f\left(\frac{5}{3}\right) + f\left(\frac{7}{3}\right) + f(3) + f\left(\frac{11}{3}\right) + f\left(\frac{13}{3}\right) + f\left(\frac{15}{3}\right)\right] \cdot \frac{2}{3} \approx 17.896\,.$$

As you should be able to tell by the figure, this is an overestimate of the actual area.

MRAM

For variety, a completely different function will be used for the midpoint rectangular approximation method (MRAM). MRAM_3 will be used to estimate the area under $g(x) = 1 + 2^{2-x}$ on the interval $[1,4]$.

Figure 13-5

The sum of the areas is $\text{MRAM}_3 = [g(1.5) + g(2.5) + g(3.5)] \cdot 1 \approx 5.475$. Visually, it is more difficult to tell whether the estimate of the area that results is too large or too small, but in this case, that is not a concern.

It is a common assumption that the average of the right rectangular approximation and the left rectangular approximation for a given number of triangles equals the midpoint approximation. This assumption is guaranteed only for linear functions. It is *not* necessarily true for any nonlinear function.

Example 13-1 illustrates the idea that MRAM is not guaranteed to equal to the average of LRAM and RRAM for a nonlinear function. No drawing will be provided for this example. For practice, sketch the function and try drawing the rectangles yourself. If you are not sure what the function looks like, graph it on your graphing calculator.

EXAMPLE 13-1

Let $p(x) = 2\ln(x)$ on the interval [3,5]. Find LRAM_4, RRAM_4, and MRAM_4 accurate to three decimal places.

In each case, $\Delta x = \frac{5-3}{4} = \frac{1}{2}$.

For LRAM, start on the left-hand edge and increase the function input by Δx each time.

$$\text{LRAM}_4 = [p(3) + p(3.5) + p(4) + p(4.5)] \cdot \frac{1}{2} \approx 5.242$$

For RRAM, do not use the left-hand endpoint. Start Δx to the right of $x = 3$.

$$\text{RRAM}_4 = [p(3.5) + p(4) + p(4.5) + p(5)] \cdot \frac{1}{2} \approx 5.753$$

For MRAM, start at the midpoint of the base of the first rectangle and increase the function input by Δx each time.

$$\text{MRAM}_4 = [p(3.25) + p(3.75) + p(4.25) + p(4.75)] \cdot \frac{1}{2} \approx 5.505$$

Note: The average of LRAM and RRAM is about 5.498, which is not equal to MRAM.

The Definition of a Definite Integral

Riemann concluded that if many triangles provide a good estimate of the area under a curve and above the x-axis, then even more triangles will give an even better estimate. Taking this to its logical conclusion, as the number of rectangles increases without bound, the error in the approximation will disappear, and the exact area will result. Historically, Riemann worked with rectangles of unequal widths, but again, the process is more easily understood in introductory calculus by using rectangles of equal width. The groundwork laid with Riemann sums should help you follow the generalization of the process, which leads to an amazing definition that changed the entire world of calculus!

Figure 13-6

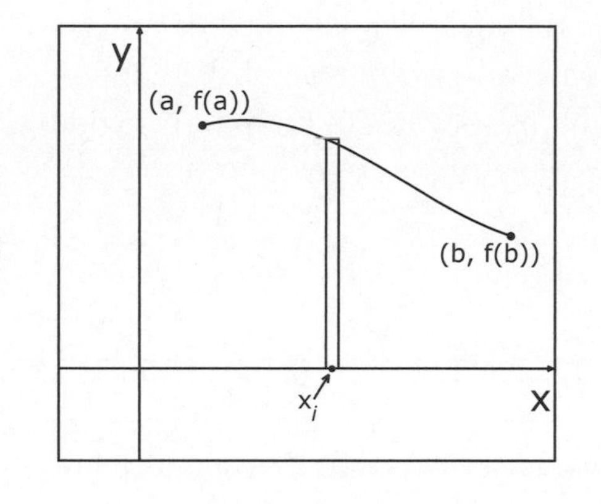

Figure 13-6 shows the graph of $f(x)$ and one sample rectangle. $f(x)$ can be any continuous function on the interval $[a,b]$. Divide the interval into n subintervals with rectangles running from $f(x)$ to the x-axis. Then $\Delta x = \frac{b-a}{n}$. The height of each rectangle can be determined by the function evaluated at any x value in each subinterval. The area of the first rectangle is denoted $f(x_1)\cdot\Delta x$, where x_1 is any x value in the first subinterval from a to $a+\Delta x$. The area of the second rectangle is denoted $f(x_2)\cdot\Delta x$, where x_2 is any x value in the interval $a+\Delta x$ to $a+2\Delta x$. Most generally, the area of the ith rectangle is $f(x_i)\cdot\Delta x$ for some x_i in the interval $a+(i-1)\Delta x$ to $a+i\cdot\Delta x$. The reason why the x_i that determines the height can be arbitrary is that as the number of rectangles increases, their width approaches zero, and the function values in the interval all converge to the same value. The sum would look like $[f(x_1)+f(x_2)+\cdots+f(x_i)+\cdots+f(x_{n-1})+f(x_n)]\cdot\Delta x$. If you look closely, this is similar to what we did a few moments ago with known numerical values and specifications on which x value to use.

The achieved sum can be written much more efficiently using summation notation. The subscript just needs to run from 1 to n.

$$[f(x_1)+f(x_2)+\cdots+f(x_i)+\cdots+f(x_{n-1})+f(x_n)]\cdot\Delta x=\sum_{i=1}^{n}f(x_i)\cdot\Delta x$$

Currently, n represents a finite amount of rectangles, so there will still be some error in the approximation. The connection to the foundation of calculus solves this issue. Riemann examined what happens to the sum as the number of rectangles increases without bound. In so doing, he defined what is known today as the definite integral.

On a closed interval $[a,b]$, $\lim_{n\to\infty}\sum_{i=1}^{n}f(x_i)\cdot\Delta x=\int_a^b f(x)dx$.

The right-hand side of the equation is read, "the definite integral of f of x from a to b." Here a and b are called the limits of integration. $f(x)$ is the integrand, and dx denotes the variable of integration.

Much more will be said about integrals in the pages ahead, but for now, please keep one key conceptual point in mind. An integral is the sum of an infinite number of products of widths and heights. What is most important to glean from this section is recognition that the limit of a Riemann sum, as the number of rectangles increases without bound, becomes a definite integral.

EXAMPLE 13-2

Write the definite integral represented by the Riemann sum

$\lim_{n\to\infty}\sum_{i=1}^{n}[3(x_i)^2+\sin(x_i)]\Delta x$ on the interval $[5,9]$.

The interval provides the limits 5 and 9. The function is in the brackets. The Δx will become dx.

$$\int_5^9[3x^2+\sin(x)]dx$$

The Trapezoidal Rule

As with all ideas in mathematics, variations on Riemann's methodology were investigated. Instead of using rectangles to approximate the area under graphs of curves, other mathematicians decided that trapezoids would provide a better approximation. The formula for the area is not quite as simple as that for the area of a rectangle, but it turns out to be very manageable. Do you remember the formula for the area of a trapezoid? A trapezoid has an upper and a lower base and a height determined by the perpendicular distance between the parallel bases. The area is half the height multiplied by the sum of the bases—that is, $A = \frac{h}{2}(b_1 + b_2)$.

To produce the trapezoids between a curve and the x-axis, the working interval is divided into subintervals, just as with the rectangles. The function values at each end of the interval are used for the lengths of the bases. Conveniently, Δx is the height of each trapezoid, because the x-axis is perpendicular to each base. Figure 13-7 shows several trapezoids inscribed under a function, $g(x)$.

Figure 13-7

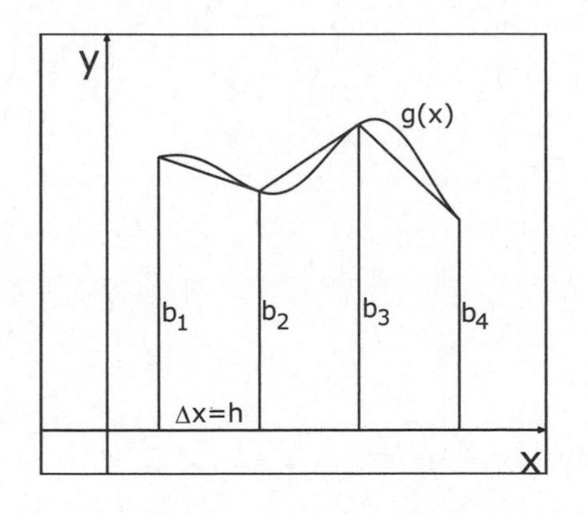

The area of the three trapezoids is

$$\frac{\Delta x}{2}[b_1 + b_2] + \frac{\Delta x}{2}[b_2 + b_3] + \frac{\Delta x}{2}[b_3 + b_4].$$

Factor out the $\frac{\Delta x}{2}$, and the area expression becomes

$$\frac{\Delta x}{2}\left([b_1 + b_2] + [b_2 + b_3] + [b_3 + b_4]\right), \text{ or } \frac{\Delta x}{2}(b_1 + 2b_2 + 2b_3 + b_4).$$

Note that the interior bases ended up being used twice, because they are shared by two trapezoids.

Trapezoidal Rule: For any given continuous function $f(x)$ on an interval $[a,b]$, the definite integral $\int_a^b f(x)\,dx$ approximated by the trapezoidal method using n trapezoids is $\int_a^b f(x)\,dx \approx \frac{\Delta x}{2}[f(x_1)+2f(x_2)+2f(x_3)+\cdots+2f(x_{n-1})+f(x_n)]$, where $\Delta x = \frac{b-a}{n}$ and each x_i differs by Δx.

Study Example 13-3 to see how the trapezoidal rule applies to a known function.

EXAMPLE 13-3

Given that $p(x) = -\frac{1}{4}x^2 + x + 3$, approximate $\int_{-1}^{3} p(x)\,dx$ on the interval [–1,3] using six trapezoids. The first and fourth trapezoids are shown in Figure 13-8 to help you visualize the situation better.

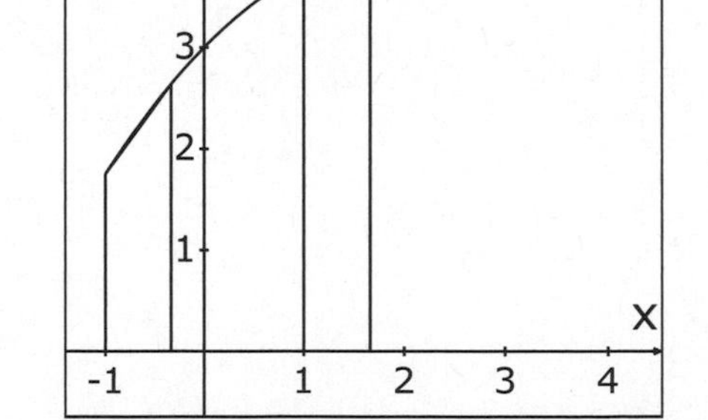

Figure 13-8

$$\Delta x = \frac{3-(-1)}{6} = \frac{2}{3}, \text{ so } \frac{\Delta x}{2} = \frac{1}{3}.$$

$$\int_{-1}^{3} p(x)\,dx \approx \frac{1}{3}\left[p(-1) + 2p\left(-\frac{1}{3}\right) + 2p\left(\frac{1}{3}\right) + 2p(1) + 2p\left(\frac{5}{3}\right) + 2p\left(\frac{7}{3}\right) + p(3)\right] \approx 13.593$$

Note that each function is evaluated at values of x that differ by $\Delta x = \frac{2}{3}$. Note also that the interior function values are doubled, because they represent lengths of bases share by two trapezoids.

The actual value of $\int_{-1}^{3} p(x)\,dx$ is $13.\bar{6}$, so the result achieved with just six trapezoids is quite good.

We have already confirmed that the average of RRAM and LRAM is only guaranteed to equal MRAM for linear functions. Amazingly, the average of RRAM and LRAM turns out to be equivalent to the trapezoidal approximation using the same number of subdivisions. This is not too difficult to confirm.

EXAMPLE 13-4

Let $f(x)$ be a continuous function on the interval $[a,b]$. Show that the average of LRAM_n and RRAM_n equals the trapezoidal approximation using n trapezoids.

Let $x_0 = a$, $x_n = b$, and $\Delta x = \frac{b-a}{n}$.

$$\text{LRAM}_n = f(x_0)\Delta x + f(x_1)\Delta x + f(x_2)\Delta x + f(x_3)\Delta x + \cdots + f(x_{n-1})\Delta x$$

$$\text{RRAM}_n = f(x_1)\Delta x + f(x_2)\Delta x + f(x_3)\Delta x + \cdots + f(x_{n-1})\Delta x + f(x_n)\Delta x$$

$$\text{LRAM}_n + \text{RRAM}_n = \Delta x[f(x_0) + 2f(x_1) + 2f(x_2) + 2f(x_3) + \cdots + 2f(x_{n-1}) + f(x_n)]$$

$$\frac{1}{2}(\text{LRAM}_n + \text{RRAM}_n) = \frac{\Delta x}{2}[f(x_0) + 2f(x_1) + 2f(x_2) + 2f(x_3) + \cdots + 2f(x_{n-1}) + f(x_n)]$$

Simpson's Rule

There is one more numerical method that gets less attention than rectangles and trapezoids, but it is actually one of the most accurate of the methods. It has been attributed to the British mathematician Thomas Simpson, but it is now commonly accepted that this method was in use previously in other parts of the world. The rule is based on a numerical method, which finds the exact area under any portion of a quadratic function by using just three equally spaced points that create two sections. It fits parabolic curves to the function, rather than the lines used by the rectangular and trapezoidal methods.

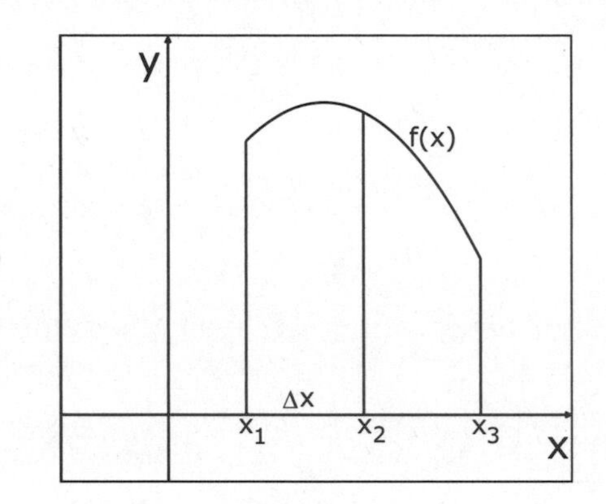

Figure 13-9

Under any positive quadratic function such as that shown in Figure 13-9, $\text{Area} = \frac{\Delta x}{3}[f(x_1) + 4f(x_2) + f(x_3)]$. Now consider applying this to a function, call it $r(x)$, for which the approximation is not exact. Just as with rectangles and trapezoids, error is reduced by using a larger number of subdivisions. To help you understand the resulting formula for Simpson's rule, picture abutting two regions under $r(x)$, each using two subdivisions. Naturally, with a total of four subdivisions, Δx will change, but the important pattern to see is what happens with the coefficients. The right-hand edge of the first region is also the left-hand edge of the second region, so $f(x_3)$ is used twice.

$$\text{Area} = \frac{\Delta x}{3}[f(x_1) + 4f(x_2) + f(x_3)] + \frac{\Delta x}{3}[f(x_3) + 4f(x_4) + f(x_5)]$$

If one more region is added, $f(x_5)$ will end the second region and begin the third, thereby being used twice.

$$\text{Area} = \frac{\Delta x}{3}[f(x_1) + 4f(x_2) + 2f(x_3) + 4f(x_4) + f(x_5)] + \frac{\Delta x}{3}[f(x_5) + 4f(x_6) + f(x_7)]$$

The result is using the first and last height once and creating an alternating pattern of coefficients of 4 and 2 for all heights between the first and the last. Note that because one section using Simpson's rule creates two divisions of the interval, the number of total subdivisions is always required to be even.

Simpson's Rule: For a continuous function $f(x)$ on an interval $[a,b]$, the approximated value of the definite integral using Simpson's rule with n subintervals, where n is a positive even integer, is

$$\int_a^b f(x)dx \approx \frac{\Delta x}{3}[f(x_0)+4f(x_1)+2f(x_2)+4f(x_3)+\cdots+2f(x_{n-2})+4f(x_{n-1})+f(x_n)]$$

where $\Delta x = \frac{b-a}{n}$.

Example 13-5 illustrates the use of Simpson's rule and demonstrates its efficiency. The result achieved with just four subdivisions is already accurate to within one-thousandth of a unit.

EXAMPLE 13-5

Given $f(x)=3+\sin(x)$, approximate $\int_1^3 f(x)dx$ using Simpson's rule and four subdivisions.

$$\Delta x = \frac{3-1}{4} = \frac{1}{2}$$

$$\int_1^3 [3+\sin(x)]dx \approx \frac{1/2}{3}\left[f(1)+4f\left(\frac{3}{2}\right)+2f(2)+4f\left(\frac{5}{2}\right)+f(3)\right]$$

$$\int_1^3 [3+\sin(x)]dx \approx \frac{1}{6}[3.841+4(3.998)+2(2.909)+4(3.599)+3.141] \approx 7.531$$

Integrals on a Graphing Calculator

Graphing calculators can be used to evaluate definite integrals and provide some insight into definite integrals as well. Calculators that are not computer algebra systems use numerical methods to evaluate definite integrals. Although tolerances may vary, most graphing calculators can provide the value of a definite integral to about five decimal places.

As in Chapter 6, the descriptions here will give directions for the most common graphing calculator used in U.S. schools: the TI-84 graphing calculator. To assess a definite integral using your graphing calculator, follow the steps listed here.

1. With the cursor on the Home screen, press the "MATH" key.
2. Press "9" which pastes "fnInt(" on the Home screen
3. The syntax is the integrand, the variable of differentiation, the lower limit, the upper limit, each separated by a comma.
4. Close the parentheses and press "ENTER" to evaluate the integral.

Figure 13-10 shows what it should look like on the Home screen.

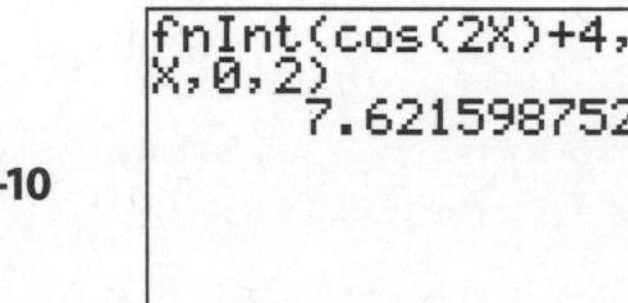

Figure 13-10

You can also evaluate a definite integral of a graph on the graphing screen. The integral evaluation tool is in the CALC menu accessed by pressing 2nd, followed by TRACE, followed by 7. You will be prompted to enter a lower limit and an upper limit. Those limits need to be values of the domain graphed on your screen. When prompted, simply type each number followed by ENTER. Figure 13-11 shows the end result of evaluating the same definite integral that was just evaluated on the Home screen.

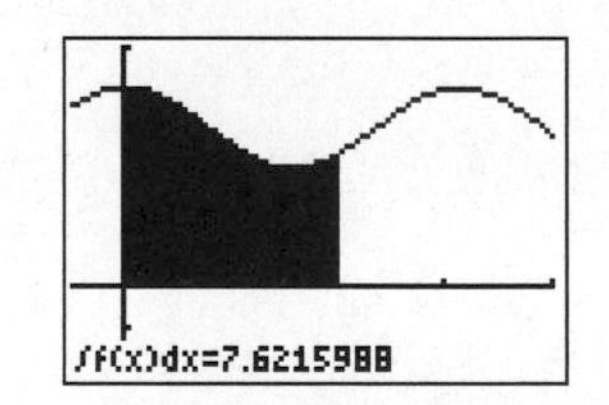

Figure 13-11

You will do much more with this tool in the following chapters.

Skill Check

Put your newfound skills to the test. Concentrate on sketching functions, rectangles, and trapezoids by hand whenever you can. Set up the approximating sums using function notation. Try to use your calculator only for evaluating the sums after you have set them up.

1. Calculate RRAM_5 for the function $h(x) = 5 + \sin\left(\frac{x}{2}\right)$ on the interval $[\pi, 2\pi]$. Use a calculator to find the answer accurate to three decimal places.

2. Calculate MRAM_4 for the function $g(x) = \sqrt[3]{x+3}$ on the interval $[0,4]$. Use a calculator to find the answer accurate to three decimal places.

3. Calculate LRAM_6 for the function $k(x) = \dfrac{1}{\ln(x)}$ on the interval $[2,5]$. Use a calculator to find the answer accurate to three decimal places.

4. A car accelerates in such a way that its speed continually increases. The velocity is measured every 3 seconds, as shown in the table. Use a left Riemann sum to determine a lower estimate for the distance traveled by the car from 3 seconds to 15 seconds.

Time (seconds)	3	6	9	12	15
Speed (feet per second)	28	50	70	88	98

5. Write the Riemann sum to represent the definite integral $\int_3^8 \frac{x}{x^2-1}\,dx$.

6. Find the area under $f(x) = \sqrt{x+5}$ on the interval $[3,9]$, using four trapezoids and four divisions of Simpson's rule. Compare your results to the area accurate to five decimal places, which is 19.83719.

7. Use your graphing calculator to evaluate $\int_{-1}^{3} (2+\sqrt{x+3})\,dx$ on the Home screen.

CHAPTER 14

The Definite Integral Explored

With any new tool, it is necessary to know how to use it properly. In mathematics, each tool is governed by a logical set of properties. Thus, it is vital to examine the properties that govern definite integrals. This chapter looks at those properties and further investigates the relationships between definite integrals and area.

Area for Negative Functions

If someone asked you for the area of your living room, you would never respond with a negative number of square feet. It is commonly accepted that area is always a positive number. This fact creates a need to be careful in the language used in connection with integrals and the area between a graph and the x-axis.

You may or may not have noticed that all functions used in the examples in the previous chapter were positive on the working domain. This guaranteed that the definite integral of the function was always positive and represented the area between the function and the x-axis. The definite integral is a Riemann sum, so on a given interval where all function values are positive, the sum of the products of function values and positive Δx values will always be positive. If a function is negative on a given integral, then the sum of the products of negative function values and positive Δx values will produce a negative value. If you want to report area, an adjustment in the result is required.

On any given interval $[a,b]$, with $h(x)<0$ over the entire interval, the area between the graph of $h(x)$ and the x-axis is $-\int_a^b h(x)dx$ to ensure that a positive value is being reported.

EXAMPLE 14-1

Use your calculator to determine the area between $p(x)=e^x-4$ and the x-axis on the interval $[-2,1]$. Give your answer accurate to three decimal places.

The graph of $p(x)=e^x-4$ lies entirely below the x-axis on the given interval, so $\text{Area}=-\int_{-2}^{1}(e^x-4)\,dx$.

Use the fnInt feature on the Home screen.

$-\text{fnInt}(e^x-4,x,-2,1)\approx 9.417$

Switching Limits

Up to this point, every Riemann sum and every definite integral has had a left-to-right orientation through the working interval. The phrasing was to "integrate from a to b." What do you think will happen if the direction of integration is reversed? Explore this with your calculator on the graphing screen. Use the positive function $g(x)=\frac{x^2}{3}+1$, and calculate the definite integral from 2 to 0 (see Figure 14-1). When your calculator prompts you for a lower limit, type a 2 even though its value is larger than 0. Then put 0 in as the upper limit.

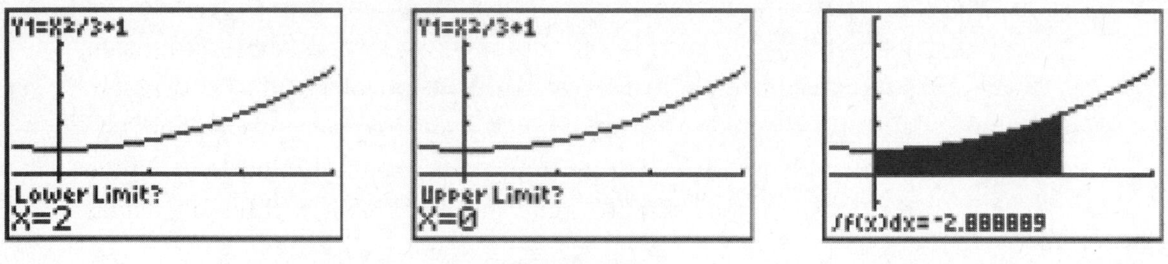

Figure 14-1

You may have predicted correctly that the result would have a negative value. Remember that the definite integral is simply the limit of an increasing number of products of function values and infinitesimally small changes in x. When the limits of the integration are reversed, moving through the interval from right to left causes Δx to be negative. The sum of the products of positive function values and negative Δx values, not surprisingly, is negative. This result is a useful property of definite integrals: $\int_a^b f(x)dx=-\int_b^a f(x)dx$. This also provides a second method for calculating the area between a negative function and the x-axis. Simply switch the limits of integration, and it will have the same effect as placing a negative sign in front of the integral.

Four More Basic Properties

A second property of definite integrals addresses constant multipliers in an integral. If the integrand has a greatest common factor that is a constant, that constant may be factored out and placed in front of the integral. There are two easy ways to understand this rule.

Because a definite integral is a limit of a summation, and you are allowed to factor constants out in front of summations and limits, the same can be done with a definite integral. For example, $\lim_{n\to\infty}\sum_{i=1}^{n}[5(x_i)^3+10x_i]\Delta x = 5\lim_{n\to\infty}\sum_{i=1}^{n}[(x_i)^3+2x_i]\Delta x$ on any interval $[a,b]$ is the same as $\int_a^b (5x^3+10x)dx = 5\int_a^b (x^3+2x)dx$.

A second connection to previous knowledge is the understanding that a common multiplier creates a vertical stretch on the graph of a function. $y=4f(x)$ just multiplies all y values of $y=f(x)$ by 4. In terms of a Riemann sum, the height of each rectangle is being multiplied by 4, while the width is not changing. Accordingly, the new result will be four times as great as the original result. In symbols, $\int_a^b 4f(x)dx = 4\int_a^b f(x)dx$.

The following example shows how each of the last two properties can be applied to simplifying and evaluating a definite integral expression.

EXAMPLE 14-2

Given that $\int_a^b h(x)dx = -8$, evaluate $\int_b^a \frac{3}{4}h(x)dx$.

$\int_b^a \frac{3}{4}h(x)dx = \frac{3}{4}\int_b^a h(x)dx$	Factor the $\frac{3}{4}$ out in front of the integral.
$\frac{3}{4}\int_b^a h(x)dx = \frac{3}{4}\cdot\left[-\int_a^b h(x)dx\right]$	Switching the limits changes the sign of the integral.
$\int_b^a \frac{3}{4}h(x)dx = \frac{3}{4}\cdot[-(-8)] = 6$	Substitute −8 for the value of the original integral.

As you move through the material, many of these basic properties will be used naturally, but it is important to lay a good foundation to build on when you encounter more challenging problems.

Under no circumstances can a variable be factored out of an integrand and placed in front of the integral. This will always lead to an incorrect answer. Only constant factors can be brought out in front of the integral.

Adding a Constant

The best way to understand how adding a constant to an integrand affects the result is to think of the transformation to the graph of the function. Consider a positive function $y = w(x)$ that has a definite integral value $\int_1^5 w(x)dx = 10$. The graph of $y = w(x) + 3$ shifts all points on $w(x)$ up 3 units. It is as if the region under $w(x)$ and above the x-axis is now sitting on a pedestal 3 units high, as shown in Figure 14-2.

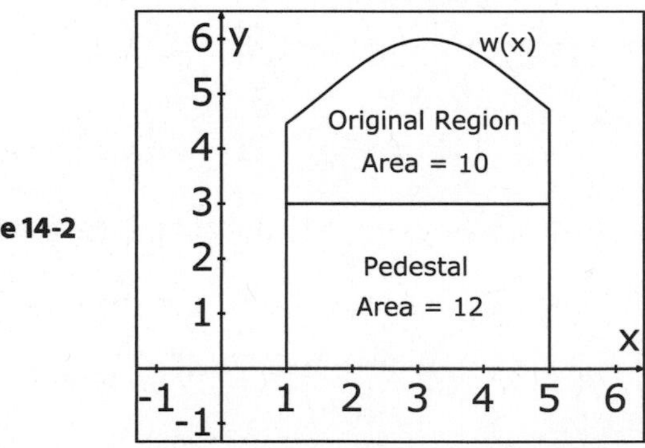

Figure 14-2

The width of the pedestal is the width of the interval, so the added area is $3(5-1) = 12$. Thus $\int_1^5 [w(x)+3]dx = 10 + 3(5-1) = 22$.

Sums and Differences

The sum and difference properties apply to integrals just as they apply to limits. That is, the integral of a sum is the sum of the integrals of the addends. In symbols, $\int_a^b [p(x)+h(x)]dx = \int_a^b p(x)dx + \int_a^b h(x)dx$. The same is true for differences.

EXAMPLE 14-3

Given $\int_2^7 f(x)dx = 8$ and $\int_7^2 h(x)dx = -11$, evaluate $\int_2^7 [3f(x)-h(x)+4]dx$.

$$\int_2^7 [3f(x)-h(x)+4]dx = \int_2^7 3f(x)dx - \int_2^7 h(x)dx + \int_2^7 4\,dx$$

$$\int_2^7 [3f(x)-h(x)+4]dx = 3\int_2^7 f(x)dx + \int_7^2 h(x)dx + \int_2^7 4\,dx$$

$$\int_2^7 [3f(x)-h(x)+4]dx = 3\cdot 8 + (-11) + 4(7-2) = 33$$

Addition of Regions

For a function $g(x)$ that is integrable over an interval $[a,c]$, the integral $\int_a^c g(x)dx$ can be written as the sum of two or more integrals over subsets of the interval. That is, for b in the interior of the interval $[a,c]$, $\int_a^c g(x)dx = \int_a^b g(x)dx + \int_b^c g(x)dx$. It does not matter whether the function changes signs in the interval.

EXAMPLE 14-4

Given $\int_1^9 g(x)dx = 17$ and $\int_1^3 g(x)dx = 7$, evaluate $\int_9^3 2g(x)dx$.

$$\int_1^3 g(x)dx + \int_3^9 g(x)dx = \int_1^9 g(x)dx$$

$$7+\int_3^9 g(x)\,dx=17$$

$$\int_3^9 g(x)\,dx=10$$

$$\int_3^9 2g(x)\,dx=20 \Rightarrow \int_9^3 2g(x)\,dx=-20$$

Net Area

When a function contains both positive and negative values in a working interval, calculating area requires the most analysis. In this situation, the value of the definite integral and the area between the graph of the function and the x-axis are not equal. The definite integral assesses regions below the x-axis as negative values, but you must count those regions as positive when you calculate the total area. In essence, for a function that takes on positive and negative values in the interval, a definite integral provides net area, not total area. For many applications of integrals, net and total areas both have significance, which will be examined further in a later chapter.

With a graphing calculator, calculating net or total area is a simple matter. The definite integral gives you the net area. In order to calculate total area, just find the integral of the absolute value of the function, and all regions of the original function below the x-axis will be calculated as positive values. Example 14-5 demonstrates the difference.

EXAMPLE 14-5

Find the net area and total area for the function $y=\cos(x^2)$ on the interval $[0,3]$.

$$\text{Net area} = \int_0^3 \cos(x^2)\,dx$$

$$\text{Total area} = \int_0^3 \left|\cos(x^2)\right|dx$$

Figure 14-3 shows the graph of $y=\cos(x^2)$ on the interval $[0,3]$. There are multiple regions above and below the x-axis.

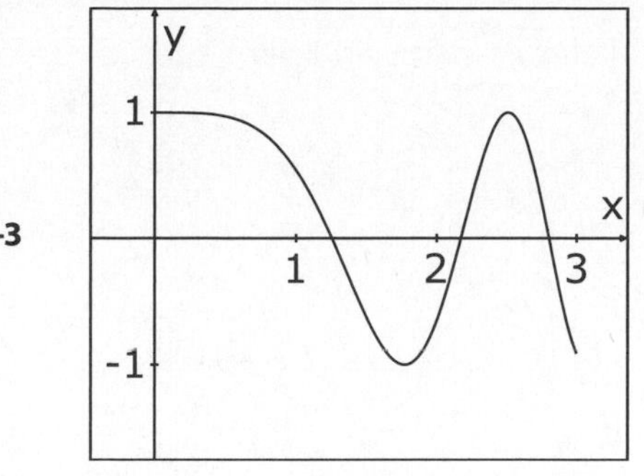

Figure 14-3

Figure 14-4 shows both calculations on the same screen.

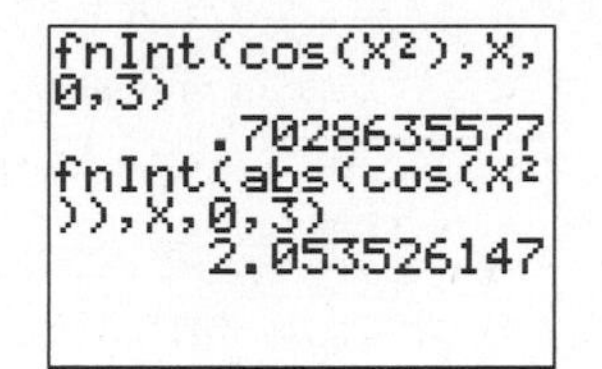

Figure 14-4

Of course, not all problems will involve using your graphing calculator. When you learn to evaluate integrals by hand, you will need to have a conceptual understanding of the difference between net area and total area. You will also have to be able to analyze where a function is positive and where it is negative.

On what part of its domain is the function $f(x) = x^2 - 2x - 8$ negative?
The factored form $f(x) = (x-4)(x+2)$ tells you the zeros are $x = -2$ and $x = 4$. Because it is a parabola that opens upward, the graph will be negative between its x-intercepts, $-2 < x < 4$.

Example 14-6 shows how mastery of the concept can also be tested without using a specific function.

EXAMPLE 14-6

In Figure 14-5, the area of each region between the graph of $h(x)$ and the x-axis is represented by a capital letter, where A, B, and C are all positive values. Simplify $\int_0^3 [h(x)+5]dx$ in terms of A, B, C, and any necessary constants.

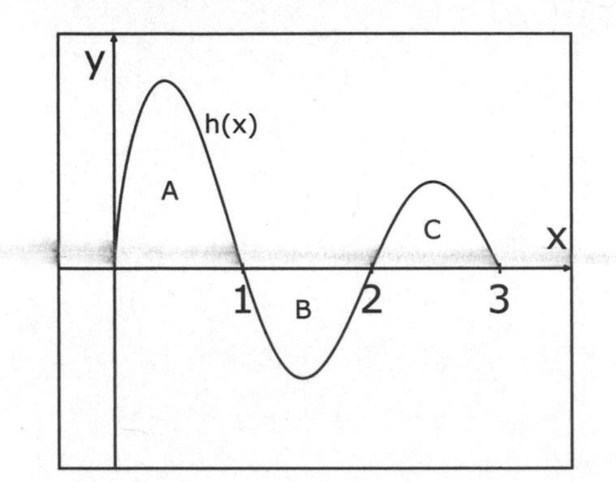

Figure 14-5

$$\int_0^3 [h(x)+5]dx = \int_0^3 h(x)dx + \int_0^3 5dx$$

The definite integral will count the area below the x-axis as negative, so the sign of B must be changed.

$$\int_0^3 h(x)dx = A - B + C \text{ and } \int_0^3 5dx = 5(3-0) = 15$$

$$\int_0^3 [h(x)+5]dx = A - B + C + 15$$

As integrals become more complicated to evaluate, the ability to break them up properly over various subintervals will be very useful!

Skill Check

With the exception of the first and last problems, the following questions are designed to test your conceptual skill without the need for a calculator. As always, give each problem your best effort before checking the solution. All questions are modeled on the topics discussed in this chapter, so going back and reviewing a section or two can also be of great assistance in mastering the skills required here.

1. Use your calculator to determine the area between $y = x^2 - 6x$ and the x-axis on the interval $[1,5]$.
2. Given that $\int_1^4 \sqrt{x}\,dx = \frac{14}{3}$, evaluate the following definite integrals, using properties you learned in this chapter.

   a. $\int_1^4 (3 + \sqrt{x})dx$ b. $\int_4^1 6\sqrt{x}\,dx$ c. $\int_4^1 (3\sqrt{x} - 5)dx$
3. On Figure 14-6, A and B are positive numbers representing the area of each region. Use A, B, and constants to write an expression for $\int_2^0 [3h(x) + 5]dx + \int_2^3 4h(x)dx$.

Figure 14-6

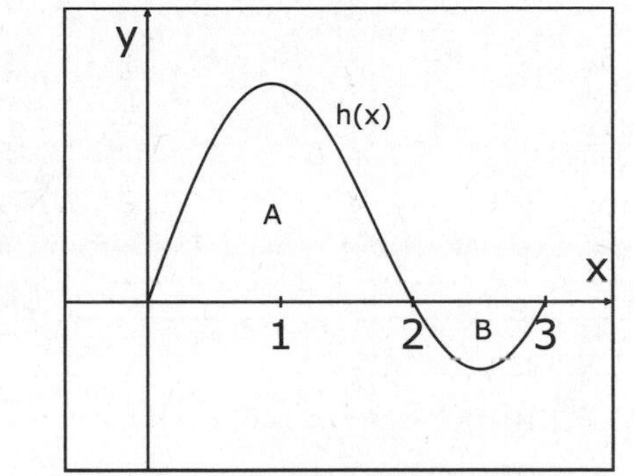

4. If $\int_a^b p(x)dx = 3a - 7b$, then find an expression in terms of a and b for $\int_b^a [p(x) + 8]dx$.
5. Use your graphing calculator to find the total area between the function $f(x) = x \cdot \sin(2x)$ and the x-axis on the interval $[0, 2\pi]$.

CHAPTER 15

The Fundamental Theorem of Calculus

With a name like the Fundamental Theorem, this must be important. In fact, it is very important! The Fundamental Theorem of Calculus has two key components and is the bridge between the two branches of the course, differential calculus and integral calculus. Its development is credited to the combined work of Isaac Newton and Gottfried Leibniz in the late seventeenth century. The concepts of this theorem became the consuming interest of many mathematicians for the next two centuries.

Integral as a Function

In order to understand the Fundamental Theorem, it is necessary to recognize a definite integral as a function of its upper limit. As you know, a function simply takes an input, acts on it, and produces an output. This is true of definite integrals, and it can be understood through exploration with a graphing calculator.

You have already evaluated a definite integral over a fixed interval with your calculator. If you hold the lower limit constant and change the upper limit, a different output is produced. In order to quickly see many cases of changing the upper limit, you can actually graph an integral with a variable upper limit!

In the "Y=" menu of your calculator, type $Y_1 = 2x+2$ and $Y_2 = \text{fnInt}(2t+2, t, 0, x)$, which is $\int_0^x (2t+2)\,dt$. The variable t is called a dummy variable, because it can be any variable, and the same result is produced. The resulting graph is a function of x. Press ZOOM, followed by 6 to get a standard window from –10 to 10. The second graph will take a moment to show up, so be patient. Your graph screen should look like Figure 15-1.

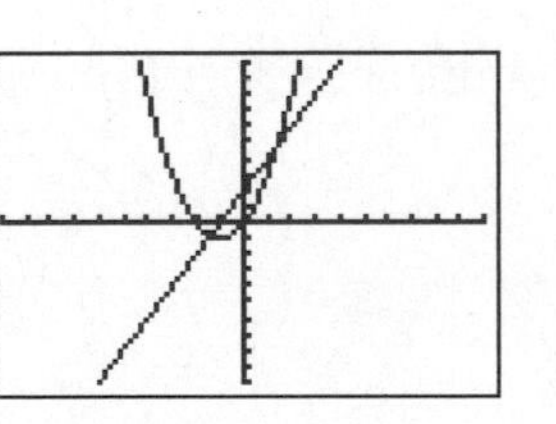

Figure 15-1

Amazingly, what your calculator just did was numerically calculate about 95 definite integrals with a lower limit of zero and an upper limit of x values between –10 and 10. The y value at each point on the parabola is a plot of an ordered pair corresponding to x as the upper limit, and y as the resulting value of the definite integral from zero to that x value. For example, the parabola contains a point (2,8), which means $\int_0^2 (2t+2)\,dt = 8$. The point (–3,3) on the parabola corresponds to $\int_0^{-3} (2t+2)\,dt = 3$. The upper limit is the changing input, and the value of the definite integral is the output, thus making the definite integral a function of its upper limit. In symbols, this can be written $\int_a^x f(t)\,dt = G(x)$ for any continuous function $f(t)$ and any constant a. The next key question is "How is the output of the integral function related to the integrand function?"

Antiderivatives

There are two ways to explore the question just posed, one more formal and one less formal. It is worth examining both. The formal method stunningly reveals how beautifully mathematics and its conclusions are interconnected. The informal method illustrates how technology has given us new ways to view old ideas.

The Formal Approach

Example 15-1 will be challenging to follow but provides a marvelous conclusion. Study it carefully.

EXAMPLE 15-1

Given $f(t) = 2t + 2$, use the limit definition of a definite integral to find a function for $\int_0^x f(t)\,dt$.

$\int_0^x f(t)\,dt = \lim_{n\to\infty} \sum_{i=1}^{n} f(x_i)\Delta x$. The independent variable, x, replaces t as the calculation proceeds through the interval $[0,x]$.

The interval is $[0,x]$, so $\Delta x = \frac{x-0}{n}$.

As you move through the interval, the height of each rectangle can be defined as $f(0 + \Delta x \cdot i) = f\left(\frac{x}{n} i\right)$.

$$\lim_{n\to\infty} \sum_{i=1}^{n} f(x_i)\Delta x = \lim_{n\to\infty} \sum_{i=1}^{n} f\left(\frac{x}{n} i\right)\Delta x$$

$$\lim_{n\to\infty}\sum_{i=1}^{n} f\left(\frac{x}{n}i\right)\Delta x = \lim_{n\to\infty}\sum_{i=1}^{n}\left[2\left(\frac{x}{n}i\right)+2\right]\left(\frac{x}{n}\right)$$

Using Gauss's formulas $\sum_{i=1}^{n} i = \frac{n(n+1)}{2}$ and $\sum_{i=1}^{n} 2 = 2n$ yields

$$\lim_{n\to\infty}\sum_{i=1}^{n}\left[2\left(\frac{x}{n}i\right)+2\right]\left(\frac{x}{n}\right) = \lim_{n\to\infty}\left[2\cdot\left(\frac{x}{n}\right)\left(\frac{n(n+1)}{2}\right)+2n\right]\left(\frac{x}{n}\right)$$

Simplifying, $\lim_{n\to\infty}\left[2\cdot\left(\frac{x}{n}\right)\left(\frac{n(n+1)}{2}\right)+2n\right]\left(\frac{x}{n}\right) = \lim_{n\to\infty}\left[\left(\frac{x^2(n^2+n)}{n^2}\right)+2x\right]$

$$\lim_{n\to\infty}\left[\left(\frac{x^2(n^2+n)}{n^2}\right)+2x\right] = x^2+2x$$

The conclusion is $\int_0^x (2t+2)\,dt = x^2+2x$. The astute calculus student will soon notice a connection between the integrand and the integral function. The mathematical beauty in this conglomeration of symbols and variables is the connection among limits, summations, ancient formulas, and an important calculus result!

The Informal Approach

The informal approach to this relationship examines a series of simple definite integrals and their resulting integral functions. Graph $\int_1^x 3\,dt$ and determine the equation of the resulting function. Next, graph $\int_{-2}^x 3\,dt$ and determine the equation of the resulting function. Determine the equation of the graph of $\int_1^x 3t^2\,dt$. Finally, do the same for $\int_{-2}^x 3t^2\,dt$.

If you were successful, the results should be as follows:

$$\int_1^x 3\,dt = 3x-3$$

$$\int_{-2}^x 3\,dt = 3x+6$$

$$\int_{1}^{x} 3t^2\,dt = x^3 - 1$$

$$\int_{-2}^{x} 3t^2\,dt = x^3 + 8$$

If you need more help to see the pattern, try taking the derivative of the right-hand side of each equation. With the exception of the change of variable, the derivative of the right-hand side of the equation is the integrand! For this reason, the resulting integral function is called an antiderivative. Another important pattern is that the constant for each antiderivative changes as the lower limit is changed. More will be said about this later.

If f is a continuous function and a is a constant, then $\int_{a}^{x} f(t)\,dt = G(x) + C$, where $G(x)$ is the antiderivative of f, and C is a constant dependent on a.

You will soon learn how to determine C, but for now, study Examples 15-2 and 15-3 to get some practice finding antiderivatives.

EXAMPLE 15-2

Determine an antiderivative of $3x^2 - 4x + 5$.

Before rules are formalized, the key is to examine the function and ask yourself, "What function has a derivative $3x^2 - 4x + 5$?"

Because powers are reduced in differentiating, start by increasing each power and expect to find a coefficient.

$$\_\_x^3 - \_\_x^2 + \_\_x^1$$

If you take the derivative of the expression above, what coefficient for each term will produce 3, 4, and 5, respectively?

The derivative of $1x^3 - 2x^2 + 5x + C$ will produce $3x^2 - 4x + 5$.

EXAMPLE 15-3

Determine an antiderivative of $x^4 + 6\cos(2x)$.

The derivative of x^5 is $5x^4$, and the derivative of $\sin(2x)$ is $2\cos(2x)$.

The original function does not have a 5 on the x^4, so place a $\frac{1}{5}$ in front of x^5 to cancel the 5.

The original function has a 6 in front of $\cos(2x)$, so place a 3 in front of $\sin(2x)$ to multiply with the chain rule factor of 2.

The antiderivative of $x^4 + 6\cos(2x)$ is $\frac{1}{5}x^5 + 3\sin(2x) + C$. You can always check an answer by taking its derivative.

The better you know your derivative rules, the easier it will be for you to find antiderivatives. If necessary, go back a few chapters and study the derivative rules again.

Rate of Change of an Integral

Because a definite integral is a function of its upper limit, it can be treated and examined just as any other function can. One of the first calculus concerns is determining the rate of change of the function, and with the knowledge you have just acquired, that task is very manageable. You know that $\int_a^x f(t)\,dt = G(x) + C$, where $G(x)$ is the antiderivative of f, and C is a constant. The rate of change of any function is its derivative. Taking the derivative of both sides of the previous equation produces $\frac{d}{dx}\left[\int_a^x f(t)\,dt\right] = \frac{d}{dx}[G(x) + C]$. But $G(x)$ is the antiderivative of f, so $\frac{d}{dx}[G(x)] = f(x)$. Because the derivative of a constant is zero, the result is the first part of the Fundamental Theorem of Calculus.

Let f be any continuous function, and let a be a constant. Then $\frac{d}{dx}\left[\int_a^x f(t)dt\right]=f(x)$. This is the first part of the Fundamental Theorem of Calculus. It asserts that the instantaneous rate of change of the integral of any function is the function itself.

This simple conclusion forged the connection between the two main branches of early calculus: differential calculus and integral calculus. It was essentially the breakthrough that the mathematics world needed to make proper use of definite integrals and all of the advantages they had to offer. It also had many implications for understanding integrals and the accumulations of rate functions. The actual task of taking a derivative of an integral is simple, as Example 15-4 illustrates, but as you will soon learn, true understanding lies much deeper.

EXAMPLE 15-4

If $y=\int_3^x \tan(t^2)dt$, find $\frac{dy}{dx}$.

By the Fundamental Theorem, $\frac{dy}{dx}=\frac{d}{dx}\int_3^x \tan(t^2)dt=\tan(x^2)$.

You do not currently know the antiderivative of $\tan(t^2)$, but if you did, the integral would produce the antiderivative with an added constant, and then the derivative of that result would take you right back to $\tan(x^2)$.

An important conceptual application of the Fundamental Theorem is used in the analysis of the rate of change of the net area under a function. When a function is defined as a definite integral, the most important first step is to write down a symbolic representation of its first derivative and, if necessary, of its second derivative. Doing so clarifies the key relationships between the functions involved in the problem. Study how this is done in Example 15-5.

EXAMPLE 15-5

Figure 15-2 is the graph of $g(t)$ and $f(x)=\int_1^x g(t)\,dt$. On the interval [–3,3], determine the intervals where $f(x)$ is increasing and where it is decreasing.

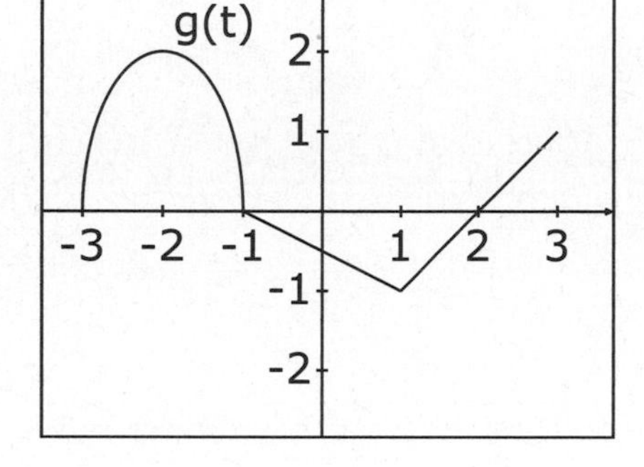

Figure 15-2

Because $f(x)=\int_1^x g(t)\,dt$, $f'(x)=g(x)$. The important connection is that the function values of $g(x)$ provide information about the slope of *f*.

f is increasing where $f'(x)=g(x)$ is positive. $g(x)>0$ on the intervals [–3,–1] and [2,3].

f is decreasing where $f'(x)=g(x)$ is negative. $g(x)<0$ on the intervals [–1,2].

You can also discover information about the concavity of a function by taking the next derivative. For instance, in the previous example, $f''(x)=g'(x)$, so the slope of the graph of g shows the sign of the second derivative of *f*, thereby providing information about the concavity of *f*.

Composite Functions in the Limits

As with most concepts in calculus, there are deeper layers. One such layer here is handling an upper limit, which is a composite function, not just a simple x. When this occurs, the chain rule must be used, just as with derivatives of typical functions. Example 15-6 demonstrates one method for tackling these problems.

EXAMPLE 15-6

If $y=\int_{3}^{x^3}\sqrt{5-t^2}\,dt$, find $\frac{dy}{dx}$.

Let $y=\int_{3}^{u}\sqrt{5-t^2}\,dt$ and $u=x^3$.

Now that the upper limit is simply u, differentiate with respect to u, using the Fundamental Theorem.

$\frac{dy}{du}=\sqrt{5-u^2}$ and $\frac{du}{dx}=3x^2$

By the chain rule, $\frac{dy}{dx}=\sqrt{5-u^2}\cdot 3x^2$.

Substituting for u yields $\frac{dy}{dx}=\sqrt{5-(x^3)^2}\cdot 3x^2=3x^2\sqrt{5-x^6}$.

A simple way to view the process is to think of it as replacing every t in the integrand with the upper limit. In the previous example, the dt becomes the derivative of x^3.

Functions in Both Limits

The Fundamental Theorem can be used even when there are functions in both limits. The properties of definite integrals enable you to break the integral into a sum of two manageable integrals, as shown in Example 15-7.

EXAMPLE 15-7

Find $\frac{dy}{dx}$ if $y=\int_{3x}^{x^2}e^t\,dt$.

Using the sum property yields $\int_0^{3x} e^t\,dx + \int_{3x}^{x^2} e^t\,dt = \int_0^{x^2} e^t\,dt$, so $\int_{3x}^{x^2} e^t\,dt = \int_0^{x^2} e^t\,dt - \int_0^{3x} e^t\,dx$.

$$\frac{d}{dx}\left(\int_{3x}^{x^2} e^t\,dt\right) = \frac{d}{dx}\left(\int_0^{x^2} e^t\,dt\right) - \frac{d}{dx}\left(\int_0^{3x} e^t\,dx\right)$$

$$\frac{d}{dx}\int_{3x}^{x^2} e^t\,dt = e^{(x^2)}\cdot 2x - e^{3x}\cdot 3$$

If $f(x)$ is a continuous function and $y = \int_{g(x)}^{h(x)} f(t)dt$, then $\frac{dy}{dx} = f(h(x))\cdot h'(x) - f(g(x))\cdot g'(x)$. Apply the Fundamental Theorem to the upper limit and then to the lower limit, and subtract the results. Remember the chain rule!

Evaluation of Integrals

The second part of the Fundamental Theorem plays a very different role. Up to this point, you have relied on your calculator to evaluate definite integrals, but now you will learn how to do this by hand.

Fundamental Theorem of Calculus: Let G be any antiderivative of the continuous function f on the interval $[a,b]$. Then $\int_a^b f(t)dt = G(b) - G(a)$. Find the antiderivative of the integrand evaluated at the upper limit minus the antiderivative evaluated at the lower limit.

The several ways of deriving the second part of the Fundamental Theorem differ in complexity. For efficiency, we will use the shortest method. You already know that a definite integral is a function of its upper limit, so let

$P(x)=\int_a^x f(t)dt$. You have also learned that a definite integral results in an antiderivative of the integrand, plus an arbitrary constant, C. Thus $P(x)$ is an antiderivative of f and differs from $G(x)$ by only a constant. That is, $G(x)=P(x)+C$. This means that $G(b)-G(a)=[P(b)+C]-[P(a)+C]$, or $G(b)-G(a)=P(b)-P(a)$. But $P(x)=\int_a^x f(t)dt$, so $P(b)=\int_a^b f(t)dt$ and $P(a)=\int_a^a f(t)dt$. Substituting the integrals in the previous equation produces $G(b)-G(a)=\int_a^b f(t)dt-\int_a^a f(t)dt$. Because there is no area under a function on an interval from a to a, $\int_a^a f(t)dt=0$. Therefore, $G(b)-G(a)=\int_a^b f(t)dt$.

In the previous chapter you were given $\int_1^4 \sqrt{x}\,dx=\frac{14}{3}$. Example 15-8 shows how the Fundamental Theorem confirms this fact.

EXAMPLE 15-8

Evaluate $\int_1^4 \sqrt{x}\,dx$.

$\sqrt{x}=x^{(1/2)}$, so it must be the derivative of $x^{(3/2)}$. To account for the $\frac{3}{2}$ that would be brought down, you need a $\frac{2}{3}$.

$$\int_1^4 \sqrt{x}\,dx \Rightarrow G(x)=\frac{2}{3}x^{(3/2)}$$

$$G(4)-G(1)=\frac{2}{3}\cdot 4^{(3/2)}-\frac{2}{3}\cdot 1^{(3/2)}=\frac{14}{3}$$

Skill Check

All of the following Skill Check problems are intended to be done without a calculator. If you apply the proper part of the Fundamental Theorem, each exercise should be relatively straightforward. Problems 5 and 6 are the only ones where you will need antidifferentiation. As always, the questions are closely related to the skills discussed in this chapter or in earlier chapters, so if you encounter difficulty, go back and reread the appropriate section.

1. Given that $f(x)=\int_2^x \sec(t)dx$, find $f'\left(\frac{\pi}{3}\right)$ without using antidifferentiation.
2. Given that $y=\int_3^{x^4} \frac{t}{\sqrt{7+t}}dt$, find $\frac{dy}{dx}$ without using antidifferentiation.
3. Find the equation of the line tangent to $h(x)=\int_{x^2}^{x^3} 2^{(t-1)}\,dt$ at $x=1$.
4. Figure 15-3 shows the graph of $p(t)$.

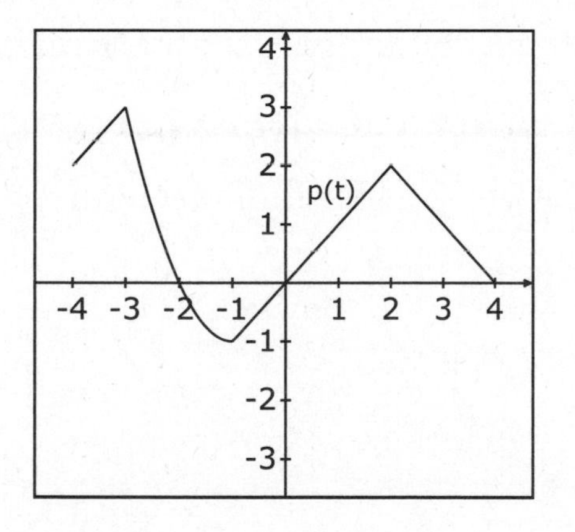

Figure 15-3

Given $r(x)=\int_{-1}^{x} p(t)dt$ on the interval $[-4,4]$, determine

a. $r(4)$
b. $r'(3)$
c. the intervals where $r(x)$ is concave down

5. Evaluate the definite integral $\int_1^2 (x^3+2x-3)dx$ by hand.
6. Evaluate the definite integral $\int_0^{\pi/2} 3\cos(x)dx$ by hand.

CHAPTER 16

Methods of Antidifferentiation

In integral calculus, the second major division of any beginning study of calculus, *antidifferentiation* is the most frequently used tool. Because antidifferentiating is essentially reversing the process of differentiation, the more familiar you are with derivative rules, the easier it will be to antidifferentiate, which is sometimes called integrating.

Geometric Methods

The first method for evaluating definite integrals is not actually a means of antidifferentiation, but it can be a powerful way to handle some integrals that might otherwise be unwieldy using analytic skills. It is also a method that is frequently overlooked—but it shouldn't be! A definite integral produces the net area between a graph and the x-axis, so using geometry to find those areas is a perfectly acceptable approach.

To use a geometric approach, you will have to be aware of the regions produced by a variety of simple functions. You will also have to be aware of when a region produces a negative value from the definite integral, and when it produces a positive value. The work you did in previous chapters with the Fundamental Theorem and with properties of definite integrals should be helpful.

Also be aware that you can make use of the properties of odd and even functions. Symmetry with respect to the origin for odd functions will often result in regions canceling each other out. This can also happen with graphs of sine and cosine. Finally, to use geometric methods, you will have to recall some simple formulas, such as those for the area of a circle, triangle, and trapezoid. Observe how this method is used in the next few examples.

EXAMPLE 16-1

Evaluate $\int_2^6 \left(\frac{1}{2}x+1\right)dx$.

The graph of the integrand is shown in Figure 16-1. Note that the area under the line and above the x-axis forms a trapezoid with the bases parallel to the y-axis. The area of a trapezoid is $A=\frac{h}{2}(b_1+b_2)$.

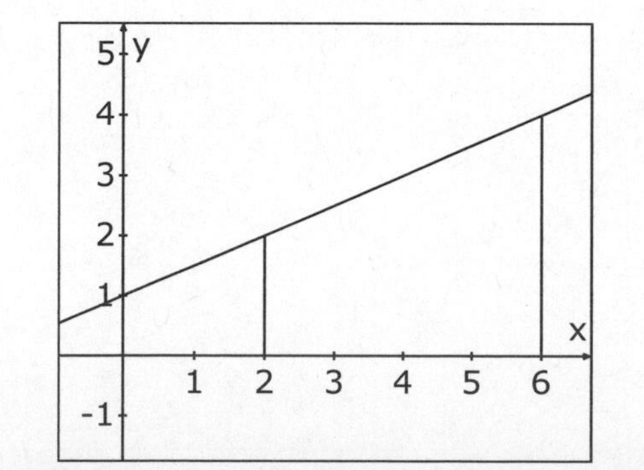

Figure 16-1

Using the function values at $x=2$ and $x=6$ as the bases,

$$\int_2^6 \left(\frac{1}{2}x+1\right)dx = \frac{4}{2}(2+4) = 12.$$

EXAMPLE 16-2

Evaluate $\int_0^3 2+\sqrt{9-x^2}\,dx$.

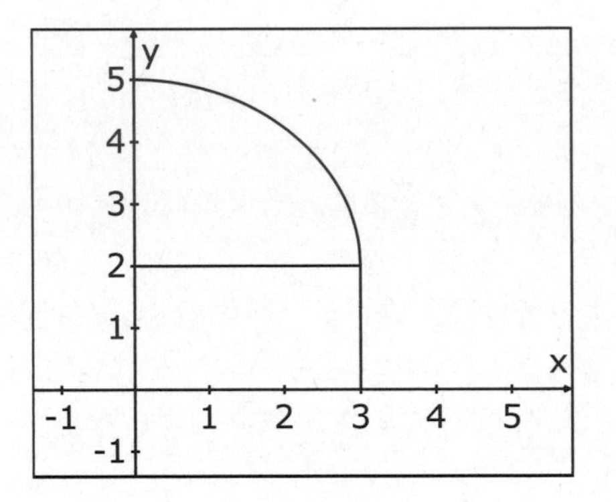

Figure 16-2

The key is to recognize $\sqrt{9-x^2}$ as the equation for a semicircle. Because $x^2+y^2=9$ is a circle, $y=\pm\sqrt{9-x^2}$ represents the upper and lower halves of the circle. With the limits from 0 to 3, the graph is just a quarter of a circle.

Note that the addition of 2 sets the quarter-circle on a pedestal 2 units high and 3 units wide.

$$\int_0^3 2+\sqrt{9-x^2}\,dx = 2\cdot 3+\frac{1}{4}\pi\cdot 3^2 = \frac{9\pi}{4}+6$$

The third example makes use of the symmetry of an odd function with respect to the origin.

EXAMPLE 16-3

Evaluate $\int_{-2}^2 x\cdot\sqrt{4-x^2}\,dx$.

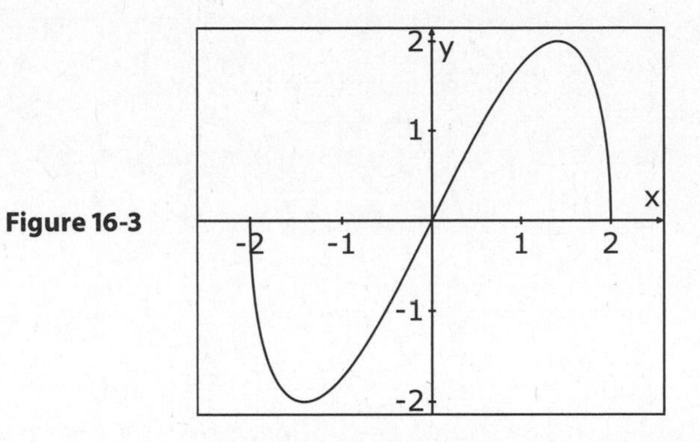

Figure 16-3

Recall that if a function $f(x)$ is odd, then $f(-x) = -f(x)$. The integrand is an odd function because replacing x with $-x$ will produce $-x \cdot \sqrt{4-x^2}$, the opposite of the original function.

Figure 16-3 shows that there is exactly the same area under the x-axis on the interval [–2,0] as there is above the x-axis on the interval [0,2]. The values cancel when you evaluate the definite integral, so $\int_{-2}^{2} x \cdot \sqrt{4-x^2}\, dx = 0$.

Even though the definite integrals in all three examples can be evaluated by hand, the geometric approach is much more efficient.

Changing the Integrand Using Algebra

Because an integral produces an antiderivative, the first goal of integration is to find the function that was differentiated to produce the integrand. Essentially, differentiation must be reversed. Sometimes that process is easy, sometimes it is more difficult, and sometimes it is just not possible.

The easiest expressions to integrate tend to be polynomials and other simple powers of variables. Recall that when you differentiate a power function, the exponent is brought down as a coefficient, and the exponent on the variable is reduced by 1. Thus, for all powers except negative one, the antiderivative is found by increasing the exponent by 1 and dividing by the new exponent.

Power Rule for Antiderivatives: Given that n is a real number not equal to negative one, the antiderivative of x^n is $\frac{x^{n+1}}{n+1}+C$. Remember that the arbitrary constant will become zero during differentiation.

The majority of the time, though, an integrand will not be so simple, and you will need to use one of many integration techniques. The first of those techniques is to make some sort of algebraic change to make the integration process easier. This could mean expanding an expression, rewriting an expression, or using a trigonometric identity. The next two examples demonstrate these various options. An alternative notation will also be introduced for writing the antiderivative and the limits at which evaluation will happen. The antiderivative will be written immediately after the integral, and an evaluation bar will be written next to the antiderivative.

EXAMPLE 16-4

Evaluate $\int_1^2 \left(x+\frac{2}{x}\right)^2 dx$.

$$\int_1^2 \left(x+\frac{2}{x}\right)^2 dx = \int_1^2 \left(x^2+4+\frac{4}{x^2}\right) dx$$ Square the binomial.

$$\int_1^2 \left(x^2+4+4x^{-2}\right) dx = \frac{x^3}{3}+4x+4\cdot\frac{x^{-1}}{-1}\Bigg|_1^2$$ Use the power rule for antiderivatives.

$$\int_1^2 \left(x^2+4+4x^{-2}\right) dx = \frac{x^3}{3}+4x+4\cdot\frac{x^{-1}}{-1}\Bigg|_1^2$$

$$\frac{x^3}{3}+4x+4\cdot\frac{x^{-1}}{-1}\Bigg|_1^2 = \left(\frac{8}{3}+8-2\right)-\left(\frac{1}{3}+4-4\right)=\frac{25}{3}$$

EXAMPLE 16-5

Evaluate $\int_0^\pi [\tan(x)\cdot\cos(x)]dx$.

Before integrating, determine whether the integrand can be written in a simpler form.

$$\tan(x)\cdot\cos(x)=\frac{\sin(x)}{\cos(x)}\cdot\cos(x)=\sin(x)$$

$$\int_0^{\pi}[\tan(x)\cdot\cos(x)]dx=\int_0^{\pi}\sin(x)dx$$

Ask the question "Sin(x) is the derivative of what function?" The derivative of $\cos(x)$ is $-\sin(x)$, so $\frac{d}{dx}[-\cos(x)]=\sin(x)$.

$$\int_0^{\pi}\sin(x)dx=-\cos(x)\Big|_{x=0}^{\pi}$$

$$-\cos(x)\Big|_{x=0}^{\pi}=-[\cos(\pi)-\cos(0)]=2$$

Any kind of algebraic or trigonometric simplification that you can use to make the integrand more manageable is always recommended. Often, integrands that look very difficult can turn out to be relatively easy if you keep this in mind.

Substitution

Time spent studying the brief table of derivatives and integrals in Appendix B will definitely improve your ability to find many simple antiderivatives. Yet there are times when the integrand is complicated enough that the antiderivative is not the least bit obvious. In those situations, a method sometimes called u-substitution can be helpful.

The purpose of u-substitution is to make it easier to see the form of an integral and thereby understand the path to the antiderivative. Within the integrand, you must identify a main function that will be called u, and you must find a constant multiple of its derivative. You should also keep your choice of u pretty simple so that its derivative is simple. The setup for u-substitution makes use of differential forms of derivatives.

The correct choice of u is paramount to the success of this process. If you know your derivative rules well, your ability to see what to choose for u will dramatically improve. For instance, in the integral $\int_1^7 7x^2 \cos(x^3)\,dx$, you should choose u to be x^3 because the rest of the integrand contains a $7x^2$, which is a multiple of the derivative of x^3. In the integral $\int_2^3 \tan^3(x)\cdot\sec^2(x)\,dx$, you might be tempted to pick $u = \tan^3(x)$, but the need to use the power rule and the chain rule makes its derivative way too complicated. The right choice is simply $u = \tan(x)$ because its derivative is $\sec^2(x)$, which appears in the integrand. Once you have successfully converted the entire integrand to a function of u, the antiderivative will usually be much easier to identify. In Example 16-3, symmetry was used to conclude that the definite integral was zero. Example 16-6 has the same integrand but the new interval eliminates the use of symmetry. Study the example to see how the substitution method is used to find the antiderivative and evaluate the integral.

EXAMPLE 16-6

Use u-substitution to evaluate $\int_0^2 x\cdot\sqrt{4-x^2}\,dx$.

Let $u = 4 - x^2$ because x is a multiple of its derivative. Then $du = -2x\,dx$ and $x\,dx = -\frac{1}{2}du$.

$$\int_0^2 x\cdot\sqrt{4-x^2}\,dx = \int_{x=0}^{x=2} -\frac{1}{2}u^{(1/2)}\,du$$ Substitute for all parts of the integrand.

$$\int_{x=0}^{x=2} -\frac{1}{2}u^{(1/2)}\,du = -\frac{1}{2}\cdot\frac{2}{3}u^{(3/2)}\Big|_{x=0}^{x=2}$$ Increase the exponent by 1 and divide by the new exponent.

$$-\frac{1}{3}(4-x^2)^{(3/2)}\Big|_{x=0}^{x=2} = -\frac{1}{3}\left[(4-2^2)^{(3/2)} - (4-0^2)^{(3/2)}\right]$$ Replace u with $4-x^2$ and evaluate.

$$-\frac{1}{3}\left[(4-2^2)^{(3/2)}-(4-0^2)^{(3/2)}\right]=-\frac{1}{3}(0-8)=\frac{8}{3}$$

Replacing Limits When Using Substitution

Note in Example 16-6 that the initial antiderivative was in terms of u, but the limits were x values. It is cumbersome to have to put the integrand back in terms of the original variable. Instead, based on your choice of u, it is much easier to replace the x-values in the limits with related values of u. Study Example 16-7 to learn how this works.

EXAMPLE 16-7

Find the value of $\int_1^2 \frac{4x+6}{x^2+3x}dx$.

Note that the numerator is a multiple of the derivative of the denominator.

Let $u=x^2+3x$ and $du=(2x+3)dx$. Thus $2du=(4x+6)dx$.

If $x=1$, then $u=1^2+3\cdot 1=4$. If $x=2$, then $u=2^2+3\cdot 2=10$.

Replace the integrand and the limits, and never return to x.

$$\int_1^2 \frac{4x+6}{x^2+3x}dx=\int_{u=4}^{u=10}\frac{2du}{u}$$

$$\int_{u=4}^{u=10}\frac{2du}{u}=2\ln(u)\Big|_4^{10}$$

$$2\ln(u)\Big|_4^{10}=2\ln(10)-2\ln(4)$$

You would get the same result if you put the antiderivative and limits back in terms of x, but doing so would be less efficient.

After substitution has happened and the integral is in the form $\int_a^b h(u)du$, why do you just get the antiderivative of $h(u)$, and not an extra factor of u for the antiderivative of du?

The du is part of a Riemann sum of an infinite number of products of widths of rectangles, du, and heights, $h(u)$, which produce the antiderivative.

Integration by Parts

There are times when direct substitution just will not work. And yet, if there are two functions in an integrand and neither is a multiple of the derivative of the other, it may still be possible to find the antiderivative using a method called integration by parts. Integration by parts is based on the process of reversing the result of a derivative found by the product rule. If you recall, for differentiable functions u and v, $\frac{d}{dx}(u \cdot v) = u \cdot v' + v \cdot u'$. This can be written in differential form as $d(uv) = u \cdot dv + v \cdot du$. Rearranging the equation and integrating both sides over any given interval produces yields $\int_a^b d(uv) - \int_a^b v \cdot du = \int_a^b u \cdot dv$, which leads to the formula for integration by parts.

Integration by Parts: If u and v are differentiable functions of an independent variable on a given interval $[a,b]$, then $\int_a^b u\,dv = uv\big|_a^b - \int_a^b v\,du$.

Integration by parts is challenging because in each problem, you must identify or find the four parts of the rule: u, dv, du, and v. This means that at times you will be differentiating and at times you will be antidifferentiating. Fortunately, there exists a reliable acronym for the priority order in which to choose u. The acronym is LIPET, which stands for **L**ogarithms, **I**nverse trigonometric functions, **P**olynomials, **E**xponential functions, and **T**rigonometric functions. Once u is identified in the integrand, all other terms comprise dv. Study Examples 16-8 and 16-9 to see how this works.

EXAMPLE 16-8

Find the area under $y = x \cdot \sin(x)$ on the interval $[0, \pi]$.

The function is all positive on the interval, which means that area is $\int_0^{\pi} x \cdot \sin(x)\,dx$.

Direct substitution will not work because neither function is the derivative of the other, so try the parts method. In the LIPET acronym, trigonometric functions are the lowest priority for u, so let $u = x$.

If $u = x$, then the rest of the integrand is $dv = \sin(x)\,dx$.

$u = x \Rightarrow du = dx$ by differentiation, and $dv = \sin(x)\,dx \Rightarrow v = -\cos(x)$ by antidifferentiation.

Use the pattern for integration by parts, $uv|_a^b - \int_a^b v \cdot du$.

$$\int_0^{\pi} x \cdot \sin(x)\,dx = -x \cdot \cos(x)|_0^{\pi} - \int_0^{\pi} -\cos(x)\,dx$$

$$-x \cdot \cos(x)|_0^{\pi} - \int_0^{\pi} -\cos(x)\,dx = [-x \cdot \cos(x) + \sin(x)]|_0^{\pi} = \pi$$

EXAMPLE 16-9

Evaluate $\int_1^3 x^2 \cdot \ln(x)\,dx$.

There are essentially only two functions in the integrand, the monomial x^2 and the natural logarithm function. Neither is the derivative of the other, so direct substitution does not apply.

A logarithm is a higher priority for the choice of u than a polynomial, so let $u = \ln(x)$. That means that $dv = x^2\,dx$.

If $u = \ln(x)$, then $du = \frac{1}{x}dx$. If $dv = x^2\,dx$, then the antiderivative is $v = \frac{1}{3}x^3$.

Using the formula for integration by parts yields
$\int_1^3 x^2 \cdot \ln(x)dx = \frac{1}{3}x^3 \ln(x)\Big|_1^3 - \int_1^3 \frac{1}{3}x^3 \cdot \frac{1}{x}dx$.

$$\frac{1}{3}x^3 \ln(x)\Big|_1^3 - \int_1^3 \frac{1}{3}x^3 \cdot \frac{1}{x}dx = \frac{1}{3}x^3 \ln(x)\Big|_1^3 - \frac{1}{3}\int_1^3 x^2\,dx$$

$$\frac{1}{3}x^3 \ln(x)\Big|_1^3 - \frac{1}{3}\int_1^3 x^2\,dx = \frac{1}{3}x^3 \ln(x)\Big|_1^3 - \frac{1}{9}x^3\Big|_1^3$$

Because the integration is the focal point of this example, the result need not be evaluated.

Additional Methods

Just when you think you have learned a lot of mathematics, you find out that the world of math is amazingly vast. And at that point, you have actually just begun to scratch the surface of all the various methods of finding antiderivatives. An entire book could be devoted to just antidifferentiation! In fact, only a minute percentage of integrals can even be evaluated analytically. Most are computed by numeric methods and approximations.

Some other means of integration, which you would encounter in a more extended course on calculus, include partial fraction decomposition, completing the square, trigonometric substitution, and reduction formulas. Before tackling more involved processes, make sure you have practiced enough to be accomplished at basic integration methods.

When you have to antidifferentiate, go through a mental checklist from easiest to most complicated method. Think first about a quick geometric approach. Next, consider simple antidifferentiation of a series of powers. If neither path leads to a solution, consider simplifying the integrand algebraically. For more complicated integrands, consider u-substitution by looking for a function and a constant multiple of its derivative. As a last resort, try integration by parts, keeping in mind the LIPET acronym for choosing what to equate with u.

Skill Check

Before working through this Skill Check, study the table of common integrals in Appendix B. When you feel prepared with the necessary facts, use the advice given in the last paragraph to attempt to evaluate the following integrals. If you want more practice with integration, the Internet is full of sites that offer practice problems. Be specific in your search. If you want practice with u-substitution, search for a phrase such as "u-substitution integration practice."

1. Evaluate $\int_3^5 \frac{x^2-4}{x-2}\,dx$.
2. Evaluate $\int_{-4}^4 -\sqrt{16-x^2}\,dx$.
3. Evaluate $\int_1^3 x^2\left(3+\frac{2}{x^3}\right)dx$.
4. Evaluate $\int_0^{2\pi} \cos(x)\,dx$.
5. Evaluate $\int_1^3 x\cdot\sqrt[3]{x^2-1}\,dx$.
6. Evaluate $\int_0^1 \frac{\tan^{-1}(x)}{1+x^2}\,dx$.
7. Evaluate $\int_1^2 x\cdot e^{2x}\,dx$.

CHAPTER 17

Indefinite Integrals

Not all integrals have to calculate an area under a curve, and the goal of antidifferentiation does not always have to be to arrive at a fixed value. Often, the goal is identifying a function to model a quantity that is changing. This is a primary role of indefinite integrals. That role may seem insignificant, but indefinite integrals are a key tool in studying a multitude of varying real-world phenomena.

Differential Equations

A *differential equation* is any equation that contains a derivative or multiple derivatives. The *order* of a differential equation is the highest derivative in the equation. For instance, $y' + y'' + x = 0$ is a second-order differential equation because it contains a second derivative. There are entire classes of differential equations—separable and nonseparable, homogeneous and nonhomogeneous, linear and nonlinear, to name only a few. An introductory calculus course focuses primarily on first- and second-order separable differential equations, because they are the easiest kind to solve and understand.

A separable differential equation is just that, able to be separated. All terms related to the dependent and independent variables may be collected, with like terms on the same side of the equation. For example, $\frac{dy}{dx} = x \cdot y$ may be rearranged into the form $\frac{dy}{y} = x\,dx$. All terms with a y are on one side of the equation, and all terms with an x are on the other side. By contrast, $\frac{dy}{dx} = x + y$ is a simple nonseparable differential equation. Solving this kind of equation is more complex and is usually taught in a course dedicated solely to differential equations.

What does it mean to solve a differential equation?

A differential equation is solved by determining the function or functions that have the differential equation as their derivative. As you might expect, solving involves antidifferentiation.

General and Specific Solutions

The two kinds of solutions to differential equations are general and specific solutions. A general solution identifies a family of functions whose derivative satisfies the differential equation. Each function in that family of functions differs only by a constant. If this sounds familiar to you, that is because we mentioned the idea of a constant difference between antiderivatives while justifying the evaluation part of the Fundamental Theorem. A specific solution identifies a single function from the family of functions. Finding a specific solution requires knowing a point contained by the solution curve. The known point is called an initial condition.

Consider the simple differential equation $\frac{dy}{dx} = x$. Without too much effort, you should be able to determine that the family of functions with this derivative is $y = \frac{1}{2}x^2 + C$, where C is any arbitrary real-number constant. This is the general solution. Figure 17-1 shows a graph of four of the curves in this family.

Figure 17-1

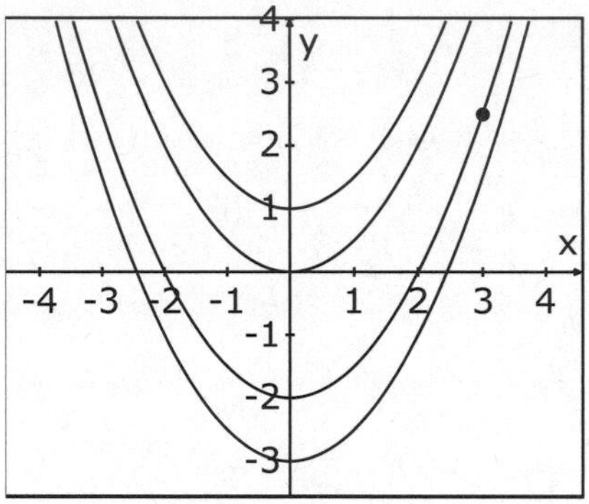

Note, however, that one of the curves has a point plotted on it. If the original task had been to find a function whose derivative was $\frac{dy}{dx} = x$ and that contained the point (3,2.5), then only one function would have satisfied those conditions. This is the specific solution to the differential equation with the given initial condition.

As always, working toward such a solution in an organized manner is preferred. In general, the process of solving a differential equation with an initial condition is as follows:

1. Separate the variables.
2. Antidifferentiate both sides of the equation, putting "+ C" on one side.
3. Substitute the initial condition into the equation.
4. Solve for the value of C and write the final equation.

Example 17-1 takes you through the process with a different differential equation.

EXAMPLE 17-1

Find a specific function such that $\frac{dy}{dx} = 3\sqrt{x-2}$ and the solution curve contains the point (6,1).

$dy = 3\sqrt{x-2}\,dx$ Separate the variables.

$\int dy = \int 3(x-2)^{(1/2)}\,dx$ Prepare to antidifferentiate.

$y = 3 \cdot \frac{2}{3}(x-2)^{(3/2)} + C$

$1 = 2(6-2)^{(3/2)} + C$ Substitute the initial condition.

$-15 = C \Rightarrow y = 2(x-2)^{(3/2)} - 15$

To check your solution, do two things. First, make sure the given point satisfies your final equation. Second, take the derivative of your solution equation and see whether it matches the original given derivative.

Slopefields

There are many differential equations for which a known symbolic solution does not exist. There are also many real-life situations which are too complex for us to fit a tidy function to them. Wind patterns and water currents are just a couple of examples. In those situations, it can still be helpful to produce a visual model of the behavior of the variables. Slopefields and (in more advanced courses) vector fields are tools that provide those visuals.

A slopefield is a visual representation of the slopes of the family of solution curves depicted at a chosen number of points. Because most functions have constantly changing slopes, the slope at each point is represented by a very short segment having that slope. There are many graphing utilities that will automatically produce a slopefield, but when you draw one by hand, it is important to understand that the slope of each segment can just be estimated. Even though a slopefield may look like a jumble of segments all over the coordinate plane, the entire picture provides insight into the family of functions, whether or not those functions can actually be determined.

Even though the differential equation $\frac{dy}{dx}=x+1$ can be solved by hand to find the family of solution curves, it is a nice simple example with which to begin learning the basics of using slopefields. To produce a slopefield by hand, you calculate the slope at a variety of points and draw a short slope mark there. This particular derivative does not depend on y, so along vertical paths where x is constant, the slopes will be the same. When $x=-1$, the slope $\frac{dy}{dx}=-1+1=0$. When $x=2$, the slope $\frac{dy}{dx}=2+1=3$. This pattern continues to produce the detailed slopefield shown in Figure 17-2.

Figure 17-2

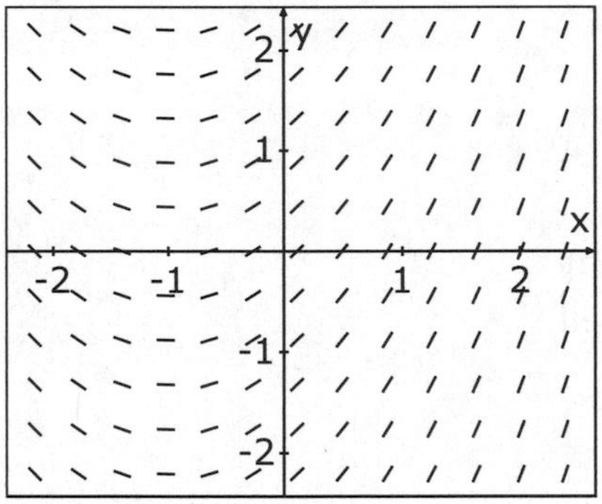

As you can see, the multitude of slope marks on the screen would be very tedious to draw by hand, but the important thing is to see the pattern of the field. The differential equation is linear, so its antiderivative is quadratic. The pattern of the slopefield shows a family of parabolas, all differing by a constant. The actual family of functions whose derivative is $\frac{dy}{dx}=x+1$ is $y=\frac{1}{2}x^2+x+C$. Figure 17-3 shows the slopefield with two possible specific solutions superimposed.

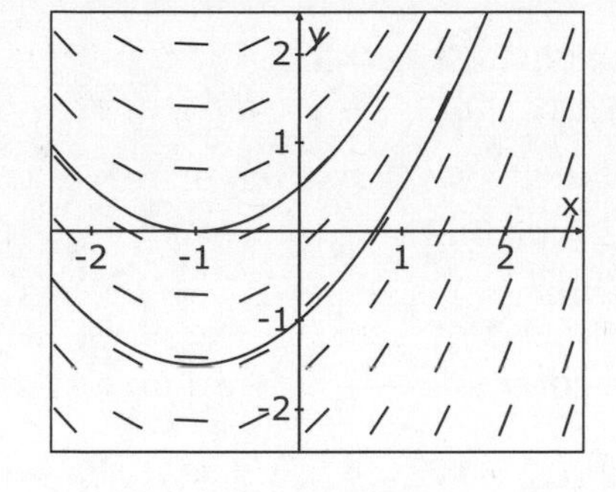

Figure 17-3

Producing a slopefield by hand by calculating the slope at one point at a time is very inefficient. Locating patterns where all slopes are the same speeds up the process significantly. In Figure 17-2, all slopes along any one vertical line remained constant because x was not changing. Try to see how skill at locating patterns can be used in Example 17-2 to produce a slopefield to help you visualize the family of solution curves when the general solution of a differential equation is not easily found.

EXAMPLE 17-2

Produce a slopefield to represent the solution curves to $\frac{dy}{dx} = \frac{x+y}{2}$.

Any path where $x+y$ remains constant will produce equal slopes. Along the line $x+y=0$, all slope marks will be horizontal. Of course, that is along the line $y=-x$.

Along the line $x+y=2$, or $y=-x+2$, all slope marks will have a slope of 1.

Recognizing this will more quickly result in the slopefield shown in Figure 17-4.

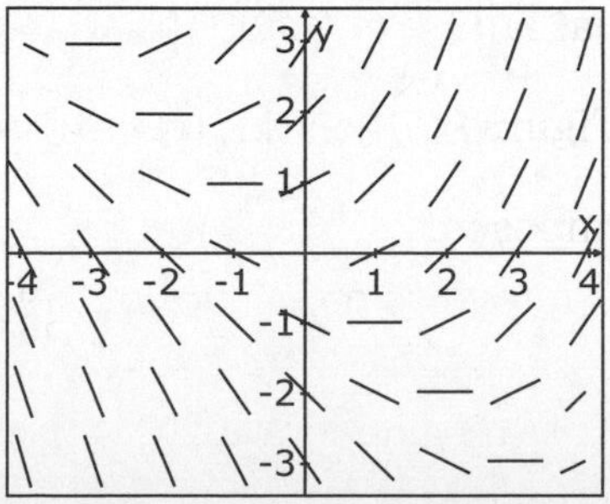

Figure 17-4

During the Gulf War, military experts studied Persian Gulf currents, essentially a slopefield, to determine where oil flows would be most likely to travel after Kuwait oil refineries were bombed. They also used wind currents to predict sand storms that would adversely affect military operations.

Exponential Growth

One of the key applications of differential equations and their solutions is in the area of exponential growth. There are a multitude of things in the world that change over varying amounts of time according to exponential models. Some of the more familiar are populations, bacterial colonies, decaying radioactive isotopes, and money. Did you know that your coffee cools according to an exponential decay model? Have you heard the saying "It takes money to make money"? When you begin investing, your money grows slowly at first. But as time passes and you accumulate more money, the rate at which it grows begins to increase. This is a simple exponential growth pattern.

A vital characteristic of exponential growth or decay models is that the rate of change of a quantity is directly proportional to the amount of the quantity present at any given time. In symbols, if Q is a quantity present as a function of time, t, and k is a constant called the constant of proportionality, then $\frac{dQ}{dt} = k \cdot Q$. Example 17-3 solves this differential equation to prove that it leads to an exponential solution.

EXAMPLE 17-3

Let Q_0 be an initial quantity present at time $t = 0$, and $\frac{dQ}{dt} = k \cdot Q$. Prove $Q = Q_0 \cdot e^{kt}$.

$\frac{dQ}{dt} = k \cdot Q \Rightarrow \frac{dQ}{Q} = k \cdot dt$ Separate the variables.

$\int \frac{dQ}{Q} = \int k \cdot dt$ Integrate both sides of the equation.

$\ln|Q| = k \cdot t + C$ Write the antiderivatives with a C on one side.

$\lvert Q \rvert = e^{kt+C}$	Rewrite the natural logarithm as an exponential equation.
$\lvert Q \rvert = e^{C} \cdot e^{kt}$	Apply properties of exponents.
$Q = Ae^{kt}$, where $A = \pm e^{C}$	
$Q_0 = Ae^0 \Rightarrow A = Q_0$	Use the initial condition, when $t = 0$, and then $Q = Q_0$.
$Q = Q_0 \cdot e^{kt}$	

As you can see, the derivation is somewhat involved, but now that it has been proved, you don't have to do the integration every time. If you're told that the rate of change of a quantity is proportional to the amount present, you can assume that it will lead to the standard exponential equation. Let's see how this understanding can be used in an application to population growth.

EXAMPLE 17-4

The rate of growth of a colony of bacteria is proportional to the amount present at any given time. The initial amount, in thousands, is 2.4. Six hours later, the colony population is 7.2 thousand. Write an equation to model the population growth, and superimpose the graph on a slope-field of the differential equation.

Start with $P = P_0 \cdot e^{kt}$ and $P_0 = 2.4$, so $P = 2.4 \cdot e^{kt}$.

Use the other information to determine k.

$7.2 = 2.4 \cdot e^{k \cdot 6}$

$3 = e^{k \cdot 6}$

$\ln(3) = \ln(e^{6k}) \Rightarrow \ln(3) = 6k$

$k \approx 0.1831$ so $P = 2.4 \cdot e^{0.1831t}$

Figure 17-5 shows the graphs of $\dfrac{dP}{dt} = 0.1831P$ and $P = 2.4 \cdot e^{0.1831t}$.

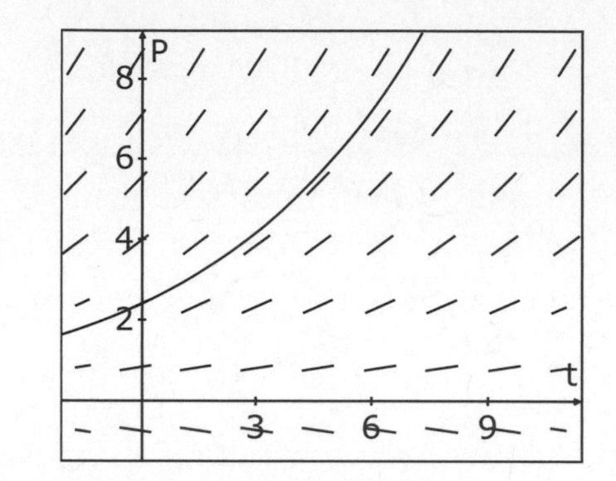

Figure 17-5

Logistic Growth

Of course, most populations do not grow exponentially forever. Many external factors influence the rate of growth over time. For example, in the 1980s, deaths due to the AIDS virus were growing exponentially. Over time, awareness and advances in treatment have significantly slowed that rate to nearly zero. Any time there are limiting factors in a potential exponential growth situation, the growth tends to follow a logistic growth curve. A plot of a simple logistic function is shown is Figure 17-6.

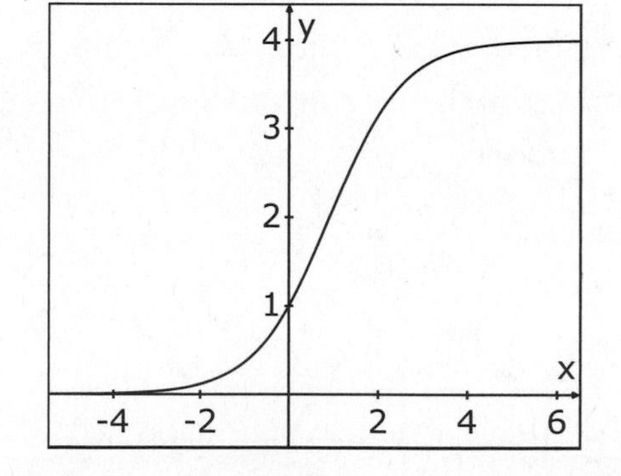

Figure 17-6

Note that the graph starts out looking exponential, but then it changes concavity and begins to level off to a maximum value, which is often called the carrying capacity for the population. The carrying capacity is the maximum sustainable population. As the population approaches its carrying capacity, the slope of the population graph approaches zero.

A logistic function is defined by a slightly varied differential equation and leads to a much more complicated solution function. Its derivation is beyond the scope of this book, but the results are valuable. If $\frac{dP}{dt} = \frac{k}{M} \cdot P(M-P)$, where k is a positive constant of growth and M is the maximum population, then $P = \frac{M}{1+Ae^{(-kt)}}$. The value of A is determined by the initial population. Note also that as the population approaches M, the value of $\frac{dP}{dt}$ approaches 0. At the introductory calculus level, you will solve these differential equations by comparison to the differential equation formula and the resulting logistic equation.

EXAMPLE 17-5

The population of bass in a lake grows according to the differential equation $\frac{dP}{dt} = 0.00006P(15000-P)$. The lake had been stocked initially with 300 bass. Find the logistic equation to model the population of the bass as a function of time.

By comparison to the equation $\frac{dP}{dt} = \frac{k}{M} \cdot P(M-P)$, $M = 15000$ and $\frac{k}{M} = 0.00006$.

$$\frac{k}{15000} = 0.00006 \Rightarrow k = 0.9$$

$$P = \frac{15000}{1+Ae^{(-0.9t)}}$$ Substitute k and M into $P = \frac{M}{1+Ae^{(-kt)}}$.

$$300 = \frac{15000}{1+Ae^{(-1.2\cdot 0)}}$$ Use the initial condition $P(0) = 300$.

$$300 = \frac{15000}{1+A} \Rightarrow A = 49$$

$$P = \frac{15000}{1+49e^{(-0.9t)}}$$

Even though solving the particular problems of exponential and logistic growth that you are encountering does not require calculus, it is important to recognize that without the calculus derivations of the equations, problems such as this might still be a mystery.

Skill Check

Attempt to do the problems in this Skill Check by hand, but recognize when it is reasonable to use a calculator. As with definite integrals, the Internet offers a myriad of examples and practice problems on these topics. You should also be able to find a free Java applet to generate slopefields from differential equations.

1. Solve the differential equation $\frac{df}{dx} = 3 + e^{1-x}$ with the initial condition $f(1) = 7$.

2. The velocity of a freely falling object as a function of time is $v = -9.8t$ meters per second. Say an object is dropped from a height of 100 meters. Find an equation for the height of the object as a function of time. Then determine how long the object will take to hit the ground.

3. Find the general solution to the differential equation $\frac{dy}{dx} = xy$.

4. Generate the marks of a slopefield for the differential equation $\frac{dy}{dx} = xy$ at the nine integer-valued points in the region $-1 \leq x \leq 1$ and $0 \leq y \leq 2$. If you have access to a slopefield generator, plot a more detailed slopefield.

5. The rate of decay of a radioactive isotope is proportional to the amount of the isotope present. If it decays 4% in the first 10 years, after how many years will only 10% of the initial amount be left?

6. A lion population is represented by the logistic differential equation $\frac{dP}{dt} = 0.72P - 0.0003P^2$, where t is measured in years. If $P(0) = 16$, write a formula for the population in terms of t.

CHAPTER 18

The Integral as an Accumulator

There are a wide variety of applications of definite integrals, and the next two chapters examine some of them. This chapter returns to definite integrals and explores their role as accumulators of quantities whose rates can be measured. Definite integrals applied as accumulators can, in much the same way as indefinite integrals, be used to extrapolate a pattern given just a rate of change and an initial condition.

Accumulation

Everyone has a sense of what accumulation means; it means collecting something over a period of time. You can accumulate wealth, or historical artifacts, or even junk around your house. Well, integrals accumulate too—not over time, but over intervals. Your experience so far should tell you integrals accumulate net area under a curve. But this tool of calculus is much more widely applicable when the curve under which area is being accumulated is the graph of a rate of change.

A definite integral accumulates that quantity whose rate of change is being integrated. If the graph of a function represents miles per hour, and the independent variable is hours, then the integral accumulates miles. If the graph of a function is the rate of change of temperature in degrees per minute, then the definite integral accumulates degrees.

You can actually watch a quantity being "accumulated" by graphing an integral on your calculator. Earlier, in Chapter 15 on the Fundamental Theorem of Calculus, you graphed an integral, but you knew less about area accumulation back then. At the time, the focus was on the integral as a function. This time, watch the values of the integral graph change, and think about the definite integral actually adding area as x increases. Set your window dimensions so that the Xmin is 0, the Xmax is 3, the Ymin is 0, and the Ymax is 8. Graph $\text{Y}_1 = -x^2 + 3x + 1$. Imagine that y represents the rate of a tank filling, in gallons per hour, and that x represents hours. If you remember a definite integral as a sum of the areas of an infinite number of rectangles under the curve, then the rectangles under this graph represent an accumulation of gallons. The rate of accumulation varies because near the ends of the parabola, the rectangles are shorter and add less area than rectangles of equal width near the vertex of the parabola. Now make $\text{Y}_2 = \text{fnInt}(-x^2 + 3x + 1, x, 0, x)$ and think about area accumulation as the integral function graphs. You can pause and restart the graphing process by hitting the ENTER button on your calculator.

Remember that the first part of the Fundamental Theorem says the rate of change of an integral is equal to the function that is being integrated. Note that the slope of the integral graph is higher where the function values of the parabola are higher. Area that represents gallons is being accumulated more quickly in those regions where the rectangles are taller.

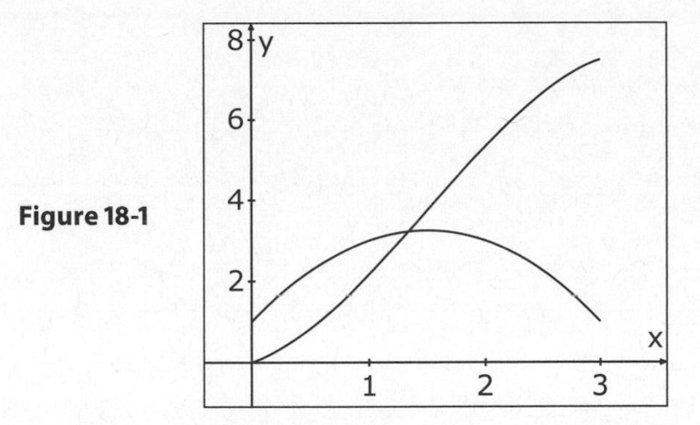

Figure 18-1

EXAMPLE 18-1

On a late summer day, the rate of change in the temperature, in degrees per hour, can be modeled by the function $y = x\sin(0.5x)$, where x is number of hours after 9 A.M. and y is the rate of change in degrees per hour. If the temperature at 9 A.M. is 68°F, then what is the temperature at 3 P.M.?

The area under this function accumulates a change in degrees. The temperature at 3 P.M. will be the starting temperature plus the accumulated degrees over the next 6 hours.

$$\text{Temperature} = 68^\circ\text{F} + \int_0^6 x\sin(0.5x)\,dx \approx 80.4^\circ\text{F}$$ Evaluate using "fnInt(".

Net Change in a Quantity

The previous section addressed strictly positive functions, and just as area under strictly positive functions is equal to total area under the curve, the same is true about total accumulation of a quantity. However, if a rate does change from positive to negative during an interval, there will be both accumulation and loss of that quantity. When this happens, the definite integral represents the net change in the quantity.

The definite integral of a rate of change, f', over a given interval $[a,b]$ produces a net change in the quantity whose rate is represented by the integrand function. In symbols, $\text{Net change} = \int_a^b f'(t)dt = f(b) - f(a)$, where f is an antiderivative of f'.

For many applications, knowing the net change in a quantity is perhaps more helpful, because when you know a beginning value of the quantity, having the net change makes it easy to calculate the ending amount.

EXAMPLE 18-2

Continuing with Example 18-1, a cold front moves in after 3 o'clock and the temperature begins dropping rapidly. The rate of change is still modeled by the same function $y = x\sin(0.5x)$, because after $x = 6$, which corresponds to 3 P.M., the function is negative, indicating a decrease in temperature. According to this model, what is the net change in temperature from 9 A.M. to 7 P.M.? And what is the temperature at 7 P.M.?

Because 7 P.M. is 10 hours after 9 A.M., integrate from 0 to 10. The net change in temperature is $\int_0^{10} x\sin(0.5x)dx \approx -9.5^\circ\text{F}$.

The actual temperature at 7 P.M. is $68^\circ\text{F} - 9.5^\circ\text{F} \approx 58.5^\circ\text{F}$.

Again, any quantity whose rate is known over time can be integrated to determine the net change in the quantity.

Average Value

Anyone with a little bit of a math background has a general idea of how to calculate average. Simply sum up a set of numbers and divide by how many numbers you summed. If this is done with function values, you might call it an *average value* of the function. The problem is that on any given interval, there are an infinite number of function values to sum. Fortunately, your experience with Riemann sums has given you the tools to deal with this dilemma!

Let $h(x)$ be a continuous function on an interval $[a,b]$. The average of n of the function values will be $\frac{1}{n}\sum_{i=1}^{n} h(x_i)$, where x_i can be any x value in each subinterval. To gain accuracy in your average, you should increase the number of values you sum by letting n increase without bound. Now your average looks like $\lim_{n\to\infty}\frac{1}{n}\sum_{i=1}^{n} h(x_i)$. But remember that $\Delta x = \frac{b-a}{n}$, so with a bit of algebra, $\frac{1}{n} = \frac{\Delta x}{b-a}$. Now your average will look like $\lim_{n\to\infty}\frac{\Delta x}{b-a}\sum_{i=1}^{n} h(x_i)$ or $\frac{1}{b-a}\lim_{n\to\infty}\sum_{i=1}^{n} h(x_i)\Delta x$. You should recall that the summation is an integral, and average value of a function is calculated by $\frac{1}{b-a}\int_a^b h(x)dx$.

An alternative way to think of an average is to distribute a quantity evenly throughout a population. For instance, if a class of twenty students pooled all of their cash and distributed it evenly, each student would receive the average amount of money. This is where the connection to net change is made. If the net area under a curve is calculated by a definite integral, when that area is divided by the width of the interval, $b-a$, the average value is the height of a single rectangle with the same area. It is as if all available area is distributed evenly over the interval. Figure 18-2 shows an example of this with a function that is entirely positive on the interval. (This rectangle concept works with functions that change signs in the interval, but choosing an all-positive function makes a first encounter with this idea is a little easier to understand.)

Figure 18-2

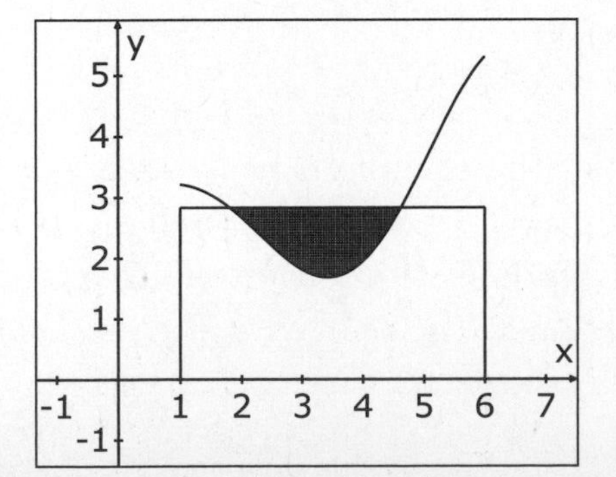

The concept of average value is also expressed as the Mean Value Theorem for Integrals, because a continuous function will always equal its average value at least once over the working interval. If $h(x)$ is continuous on $[a,b]$, then there exists some x_i in the interval such that $h(x_i)=\frac{1}{b-a}\int_a^b h(x)dx$.

The average value of the function shown on the interval [1,6] is about 2.832. A rectangle has been drawn with this height. The shaded region between the curve and the rectangle is where the rectangle overestimates the area. This region is balanced perfectly by the outer sections where the rectangle underestimates the area between the curve and the x-axis. Imagine taking the region between the curve and the x-axis and reshaping it into a rectangle of the same width. That is the average value.

Example 18-3 presents one practical application of average value.

EXAMPLE 18-3

The inventory of refrigerators in a warehouse is modeled by $I(t)=\frac{1800}{1+e^{(0.1t)}}$, where t is the number of days since the first day of the month. If it costs approximately $0.16 per refrigerator per day to store them, what is the approximate monthly cost of storing refrigerators in a typical 30-day month?

Because the inventory is constantly changing, it is best to work with the average number of refrigerators stored each day, so calculate the average value.

$$\frac{1}{30-0}\int_0^{30}\frac{1800}{1+e^{(0.1t)}}dt\approx 387$$

Thinking about the units is helpful here. $I(t)=\frac{1800}{1+e^{(0.1t)}}$ is the number of refrigerators in the warehouse on any given day. Even though it is infinitely small, dt represents a portion of days. But 30 – 0 also represents days. Thus 387 represents the average number of refrigerators in the warehouse each day of the month.

The monthly cost of storing refrigerators is about (387)(30)($0.16), or $1857.60.

Total Distance and Displacement

One of the most common uses of the integral as an accumulator is in the context of motion. Velocity is the instantaneous rate of change of position, so when velocity is integrated, the result accumulates position. The function values that determine the height of each rectangle might be in miles per hour, and values on the horizontal axis might be in hours. When the products of width and height are summed, each term of the sum is miles per hour times hours, and miles traveled are accumulated.

Of course, when working with velocity you must pay attention to the direction of motion, because velocity is a signed quantity. Traditionally, an object moving upward from the ground is considered to have positive velocity, because its distance from the ground is increasing while time is increasing. On the downward flight, velocity is negative. Similarly, for an object moving along a horizontal axis, motion to the right is usually positive velocity and motion to the left is negative. The direction and sign of velocity are significant, because a change in the sign of velocity causes total distance traveled to differ from displacement. The simplest possible example is walking forward 6 feet and backward 2 feet. You have traveled a total of 8 feet, but you are displaced only 4 feet from your initial position. A definite integral of velocity without regard for sign produces displacement, or net change in position. To calculate total distance traveled, you must calculate the total area between the velocity graph and the time axis. As you learned earlier, regions below the axis must be counted as positive.

EXAMPLE 18-4

The velocity of a particle moving along a horizontal path is defined by $v(t) = 2t - t^2$, where velocity is in feet per second and time is in seconds. Calculate the displacement and total distance traveled in the first 4 seconds.

$$\int_0^4 2t - t^2\,dt = t^2 - \frac{1}{3}t^3\Big|_0^4$$ The definite integral finds displacement.

$t^2 - \frac{1}{3}t^3\Big|_0^4 = \left(4^2 - \frac{1}{3}\cdot 4^3\right) = -\frac{16}{3}\text{ft}$ The particle ends $5\frac{1}{3}$ feet to the left of where it started.

The graph of velocity will help in determining total distance traveled.

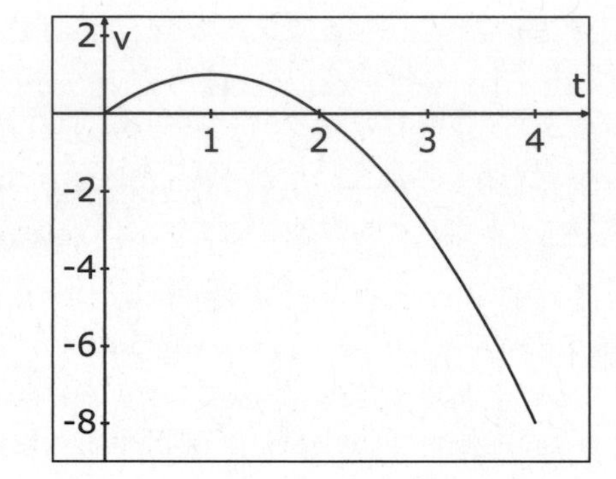

Figure 18-3

Note that velocity is positive for $0 < t < 2$ and negative for $2 < t < 4$. Two integrals are necessary, one of which treats the second interval as a positive area.

Total distance is $\int_0^2 2t - t^2\,dt + \int_4^2 2t - t^2\,dt$. The limits were switched to create a positive area on the interval [2,4].

$t^2 - \frac{1}{3}t^3\Big|_0^2 = \frac{4}{3}\text{ft}$ and $t^2 - \frac{1}{3}t^3\Big|_4^2 = \frac{20}{3}\text{ft}$ Calculated as before.

The total distance traveled is $\frac{24}{3}$ feet, or 8 feet.

You may remember from a previous chapter that with a calculator, finding the total area between a graph and the axis is very simple. Just calculate a numerical integral of the absolute value of the function.

On a given interval, when are total distanced traveled and displacement the same?

On a given interval, total distance and displacement are the same when the velocity function does not change sign in the interval. The object is always moving in one direction.

Skill Check

Unless otherwise instructed, attempt to solve the following problems without a calculator. In some cases, you will be instructed to integrate by hand and then, to avoid a lot of arithmetic gymnastics, to evaluate the result with a calculator.

1. The rate of change of motorists passing through a toll plaza is modeled by the function $c(t) = -0.7t^2 + 6t + 5$ where c is hundreds of cars per hour, and t is the number of hours since noon. Write a short explanation of the meaning of the integral $\int_3^6 c(t)\,dt = 51.9$.
2. A cup of coffee is poured at 185°F and cools at a rate defined by $\frac{dH}{dt} = -10e^{\left(\frac{-t}{13}\right)}$, where H is the temperature of the coffee, and t is minutes since the coffee was poured. Integrate by hand, and then use a calculator to evaluate to find the temperature of the coffee 5 minutes after it was poured.
3. Find the average value of $f(x) = e^x - 4x$ on the interval $[0,3]$. Antidifferentiate by hand, and then evaluate with a calculator.
4. Examine the graph of $g(x)$ in the figure, and explain whether the average value on the interval $[0,4]$ is a negative or a positive value.

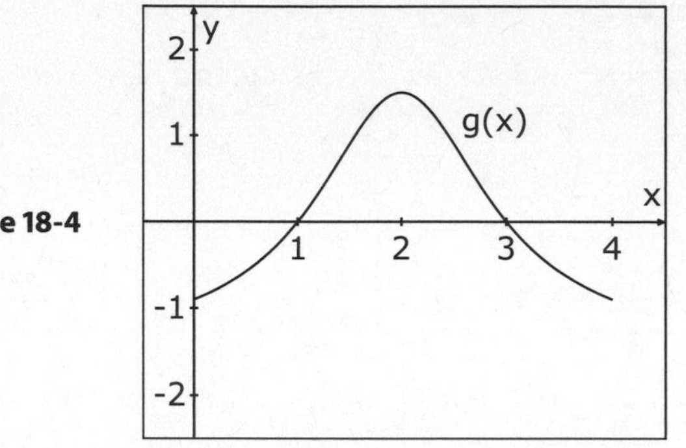

Figure 18-4

5. A particle starts at position (0,3) and moves along the y-axis with velocity defined as $v(t)=\frac{1}{2}t\cdot\sin(t^2+1)$. What is the position of the particle when $t=2.3$? Antidifferentiate by hand, and then evaluate by calculator.

6. An object moves along a horizontal path. The velocity in feet per second is $v(t)=3t^2-8t+4$. Set up an integral expression using three definite integrals to determine the total distance traveled during the first 4 seconds. Do not integrate.

CHAPTER 19

Applications of Integrals

The applications of integrals are almost unlimited, but one of their most interesting and widely practiced uses is in finding area between curves and volumes of irregularly shaped objects. Do you know how much concrete went into the construction of the Hoover Dam? In order to estimate the required materials and the resulting cost, engineers had to model the dam. They almost certainly used calculus to determine its approximate volume and the need for 3,250,000 cubic yards of concrete!

Area Between Curves

You have already spent some time studying the area between a curve and the x-axis, but there are times when the area between two curves is needed. Fortunately, if you have developed a grasp of Riemann sums, the task ahead will be very manageable. Instead of envisioning rectangles running from a graph to the x-axis, simply run the rectangles from one function to another. The definition of the height of each rectangle will then be just the difference in the function values. Figure 19-1 shows one rectangle running between two functions h and g.

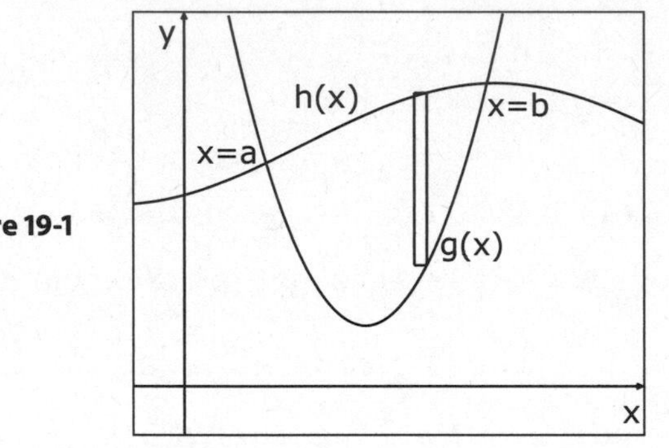

Figure 19-1

In this case, the height of each rectangle is $h(x)-g(x)$, and the width of each is still Δx. The Riemann sum is then $\lim_{n\to\infty}\sum_{i=1}^{n}[h(x_i)-g(x_i)]\Delta x=\int_a^b[h(x)-g(x)]dx$.

Even though the functions shown are both above the x-axis, it does not matter if they are below the axis or partially above and partially below. When you are finding the area between two curves, the orientation of the graphs to the x-axis does not matter. As long as the integral is set up with the upper function minus the lower function and is integrated left to right from point of intersection to point of intersection, the result will be correct.

In the working interval, if the functions switch orientation with each other, you must be careful to set up several intervals with the subtraction of the functions in the proper order. Figure 19-2 shows how two functions can switch orientation and create the need for several integrals.

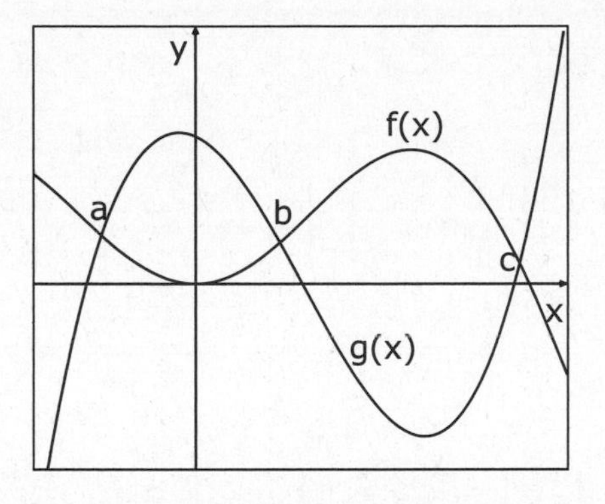

Figure 19-2

Given that the three intersections occur at $x=a$, $x=b$, and $x=c$, $g(x)$ is above $f(x)$ on the interval $[a,b]$ and below $f(x)$ on the interval $[b,c]$. Without technology, two integrals are required. The total area between the curves is $\int_a^b [g(x)-f(x)]dx + \int_b^c [f(x)-g(x)]dx$. If you *did* have a graphing calculator available for the problem, integrating the absolute value of the difference would be very efficient. It would remove any concern about which function lies above or below the other. The integral would then be $\int_a^c \left| g(x)-f(x) \right| dx$.

When you are finding area by hand, sometimes the biggest challenge is the algebra needed to find the points of intersection. Example 19-1 shows how to do an "area between curves" problem from start to finish.

EXAMPLE 19-1

Find the area of the region enclosed by $f(x)=x-2$ and $g(x)=x^2-4x+2$.

$x^2-4x+2=x-2$ Set them equal to find points of intersection.

$x^2-5x+4=0 \Rightarrow x=4$ or $x=1$ Solve by factoring.

A quick sketch will show that the line is above the parabola.

$$\int_1^4 (x-2)-(x^2-4x+2)dx = \int_1^4 (-x^2+5x-4)dx$$

$$\int_1^4 (-x^2+5x-4)\,dx = \left(-\frac{1}{3}x^3+\frac{5}{2}x-4x\right)\Bigg|_1^4$$

$$\left(-\frac{1}{3}x^3+\frac{5}{2}x-4x\right)\Bigg|_1^4 = \left(-\frac{64}{3}+10-16\right)-\left(-\frac{1}{3}+\frac{5}{2}-4\right)=\frac{9}{2}$$

ALERT

When you are working with areas and volumes, it is always important to know what the region and object look like. This topic in calculus is very visual, and working on the right region is essential to success. Knowing the graphs of the nine basic functions is paramount.

Sometimes the regions described will be determined by more than two functions. In those cases, it may be necessary to write multiple integrals to find the enclosed area. Example 19-2 shows how this is handled.

EXAMPLE 19-2

Set up an integral expression to find the area of the region R, enclosed by the functions $f(x)=\sqrt{x+1}$, $g(x)=x-1$, and $h(x)=-x+1$ (see Figure 19-3).

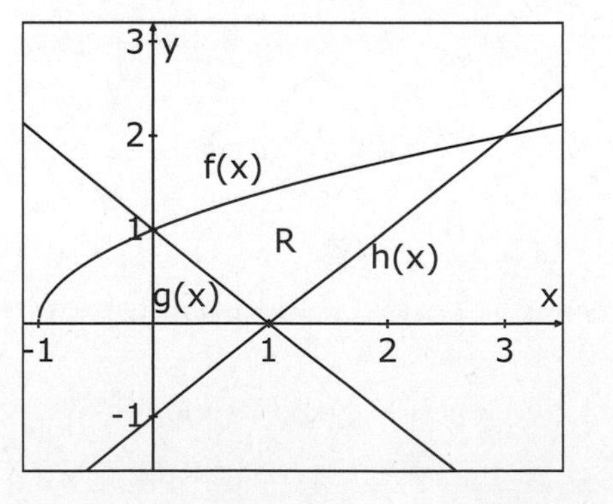

Figure 19-3

Note that for $0 \leq x \leq 1$, the vertical distance between the functions is determined by $f(x)$ and $g(x)$ and that for $1 \leq x \leq 3$, the distance is defined by $f(x)$ and $h(x)$. Two integrals can be used to find the area of each region and the results summed.

$$\int_0^1 [f(x)-g(x)]dx+\int_1^3 [f(x)-h(x)]dx$$

$$\int_0^1 [\sqrt{x+1}-(x-1)]dx+\int_1^3 [\sqrt{x+1}-(1-x)]dx$$

In completing the problem, take care to distribute the negative signs properly.

Of course, good mathematicians are always keeping watch for alternative ways to tackle challenges. Look again at Figure 19-3. Because the lower functions are linear, you could simply find the area under $f(x)=\sqrt{x+1}$ on the interval $[0,3]$ and subtract the areas of the two triangles between the lines and the x-axis.

Finding a Volume by Cross Sections

A typical loaf of bread consists of a number of individual slices, each with its own volume. Each piece is roughly the shape of a prism, with a height and two parallel faces. If you add up the individual volumes of the slices, the sum is the entire volume of the loaf. This is also true for a tomato that is cut into a series of circular slices, each with its own volume. Of course, the tomato slices are shaped more like very short cylinders with parallel circular bases and varying radii. This is the premise behind finding the volume of any solid object with cross sections that can be measured. Of course, if Riemann had been trying to find the volume of the tomato, he would have cut his slices infinitely thin!

The volume of a solid with cross-sectional areas dependent on x is found by summing the volumes of all cross sections. The volume of each cross section is the product of the cross-sectional area, call it $A(x)$, and the cross section's height, call it Δx. In symbols, $V=\lim_{n\to\infty}\sum_{i=1}^{n} A(x_i)\Delta x=\int_a^b A(x)dx$.

Understanding this principle is as easy as counting the slices in a loaf of bread, but actually producing the integrand can be very challenging. The hardest part is often visualizing the solid being described. Once that is done, and the cross-sectional area is described in terms of x, the antidifferentiation is usually straightforward. Close your eyes and try to picture a solid object whose cross sections are semicircles perpendicular to the x-axis, with their diameters in the xy-plane running between the graphs of $f(x)=\sqrt{x}$ and $g(x)=-\sqrt{x}$ on the domain $1 \le x \le 9$. The surface of the solid is shown in Figure 19-4, with just the end diameters drawn. Imagine slicing this shape with cuts perpendicular to the x-axis. Each slice is a semicircle with a radius dependent on the x value where the cut is made. The diameter is $d=\left[\sqrt{x}-(-\sqrt{x})\right]=2\sqrt{x}$, so the radius at any x_i is $r=\sqrt{x_i}$. The area of any one cross section is $A(x_i)=\frac{1}{2}\pi r^2=\frac{1}{2}\pi x_i$. The volume is the Riemann sum, $V=\lim_{n\to\infty}\sum_{i=1}^{n}A(x_i)\Delta x=\int_1^9 \frac{1}{2}\pi x\,dx$. As you can see, the resulting integral is not too difficult to evaluate.

Figure 19-4

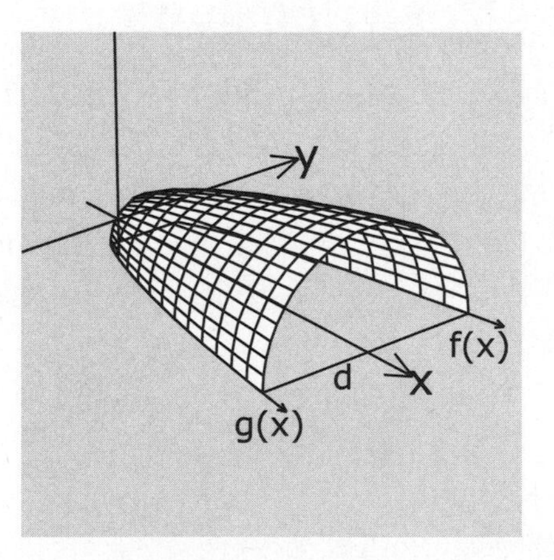

Example 19-3 will take you step by step through the process of finding a volume by cross sections.

EXAMPLE 19-3

Find the volume of a solid with cross sections that are isosceles right triangles. The hypotenuse of each triangle is a chord of the circle

$x^2 + y^2 = 4$ and is perpendicular to the x-axis. The solid is shown in Figure 19-5.

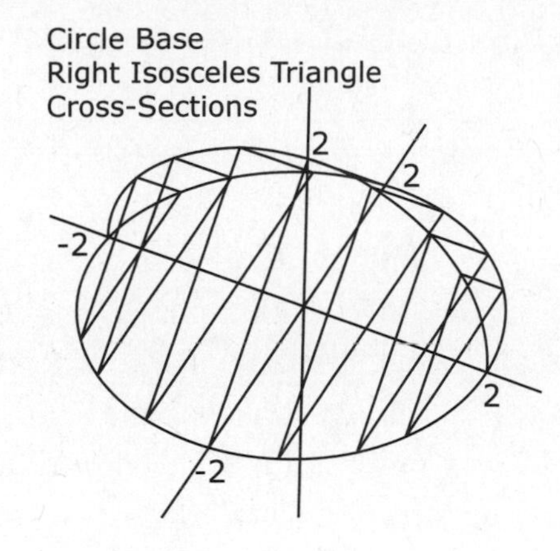

Figure 19-5

The equation of the circle is the union of two functions $y = \pm\sqrt{4 - x^2}$. This means that each hypotenuse runs from $\left(x_i, \sqrt{4 - x_i^2}\right)$ to $\left(x_i, -\sqrt{4 - x_i^2}\right)$, which makes its length $2\sqrt{4 - x^2}$.

If the length of the hypotenuse of an isosceles right triangle is $2\sqrt{4 - x^2}$, then one leg is $\frac{2\sqrt{4 - x^2}}{\sqrt{2}} = \sqrt{2} \cdot \sqrt{4 - x^2}$.

The area of an isosceles right triangle is $A = \frac{1}{2}(Leg)^2$, so $A(x) = \frac{1}{2}\left(\sqrt{2} \cdot \sqrt{4 - x^2}\right)^2 = 4 - x^2$.

$$V = \int_{-2}^{2} (4 - x^2)\,dx = \left(4x - \frac{1}{3}x^3\right)\Big|_{-2}^{2}$$

$$\left(4x - \frac{1}{3}x^3\right)\Big|_{-2}^{2} = \left(8 - \frac{8}{3}\right) - \left(-8 - \frac{-8}{3}\right) = \frac{32}{3}$$

Finding a Volume by Discs

If a three-dimensional figure can be modeled by the rotation of a curve around a line, then the method of cross sections becomes a special case where all the cross sections are circular. The thickness of Δx makes each slice a disc, and the same formula can be applied in each situation. You should remember from geometry that the volume of a cylinder is $V = \pi \cdot r^2 h$. When you are calculating volumes of revolution by discs, the varying radius is dependent on a function, and the height is $h = \Delta x$.

Suppose the simple line $y = x$ on the interval $0 \le x \le 5$ is rotated around the x-axis. Do you know what familiar shape is formed? If you guessed a cone, congratulations! The ability to visualize the result of a revolution readily is a big help. If you do struggle in attempting to visualize these figures, try to see just one rectangle rotated around, and the disc that is formed by that rectangle. If you can figure out a formula for the volume of one disc, the integral will take care of the rest. Also, there are many free online applets that can produce three-dimensional renderings to assist you in developing the skill to envision the rotated surface.

In the case of the cone, the result of that revolution is an object whose volume can be calculated without calculus. Of course, there are an unlimited number of examples of rotated functions that do require calculus. Example 19-4 considers one such function.

EXAMPLE 19-4

On the interval $[0,3]$, the region below the graph $p(x) = 3e^{-x}$ and above the x-axis is rotated around the x-axis. Set up an integral and find the volume of the solid generated. The resulting solid is shown in Figure 19-6.

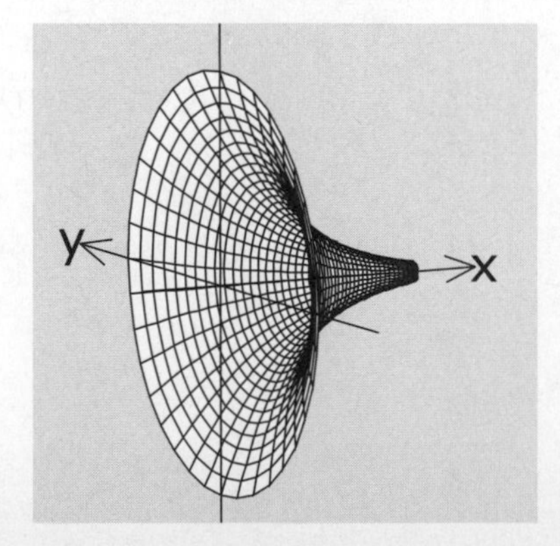

Figure 19-6

The radius of each disc varies with $p(x)$ such that $r_i = p(x_i)$ for each disc in the interval.

The volume of any one disc is $V_i = \pi(r_i)^2 h = \pi[p(x_i)]^2 \Delta x$.

The resulting Riemann sum is $\lim_{n\to\infty}\sum_{i=1}^{n}\pi[p(x_i)]^2\Delta x = \int_0^3 \pi[3e^{-x}]^2\,dx$.

$$\int_0^3 \pi[3e^{-x}]^2\,dx = \int_0^3 9\pi e^{-2x}\,dx$$

$$\int_0^3 9\pi e^{-2x}\,dx = -\frac{1}{2}\cdot 9\pi e^{-2x}\Big|_0^3$$

$$-\frac{1}{2}\cdot 9\pi e^{-2x}\Big|_0^3 = -\frac{9\pi}{2}(e^{-6} - e^0) \approx 14.102$$

With practice, you'll learn that finding a volume of revolution by the disc method boils down to defining the radius. When a region between a function $g(x)$ and the x-axis is rotated around the x-axis, the radius is always $g(x)$. If a region is rotated around a line other than the x-axis, the definition of the radius changes!

Finding a Volume by Washers

Our next topic is finding volumes of revolution when the region being rotated is not directly adjacent to the axis of rotation. When it *is* directly adjacent, the object formed is a solid and its cross sections are solid discs. But if there is space between the region rotated and the axis of rotation, then there will be empty space on the interior of the three-dimensional object created. Figure 19-7 shows an example of a "solid" of rotation with some empty space.

Figure 19-7

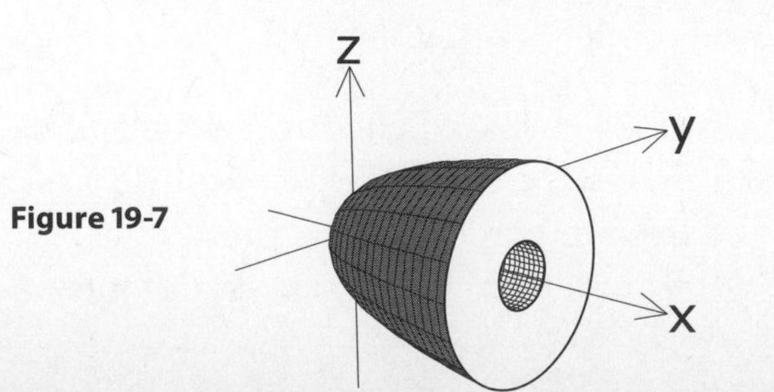

When the object is sliced in a direction perpendicular to its central axis, the slices will still be circular, but they will have holes in the middle of them. The shape of the slices is just like that of a washer, which is why the method is often referred to as finding volumes by washers. The Riemann sum is adding up the volumes of an infinite number of washers. Do you remember how to find the volume of a washer? A washer is nothing more than a short cylinder with a cylinder removed from its center. There is an outer radius of the original cylinder and an inner radius of the cylinder removed. The volume is $V = \pi R^2 h - \pi r^2 h$, where R is the outer radius and r is the inner radius. Figure 19-8 shows a generic washer with the inner and outer radii, along with the height, labeled.

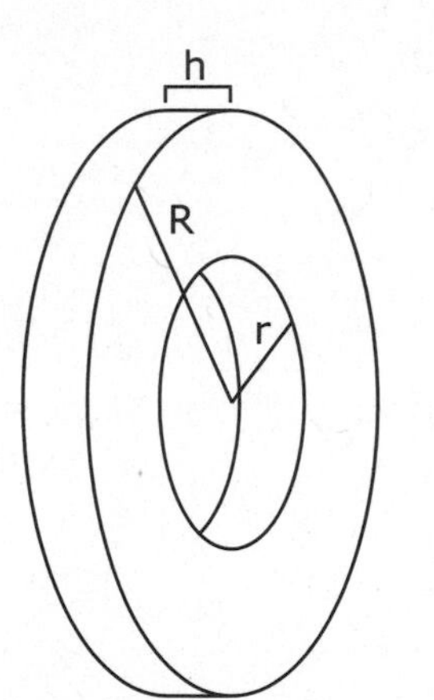

Figure 19-8

Volume by Washers: Given that two continuous functions $f(x)$ and $g(x)$ intersect at $x = a$ and $x = b$, with $b > a$ enclosing a region P *not* adjacent to the x-axis. And given that $f(x)$ is the boundary of P farther from the x-axis than $g(x)$. Then the volume of revolution when P is rotated around the x-axis is $V = \pi \int_a^b [f(x)^2 - g(x)^2] dx$.

When functions are being rotated, the challenge in setting up the proper integral is defining each radius correctly. It is important to realize that the function farther from the axis of rotation determines the outer radius, and that the function closer to the axis of rotation determines the inner radius. Just as in using the disc method, visualizing the washer created by just one rectangle is often enough for you to understand the problem. Example 19-5 provides a first look at this method.

EXAMPLE 19-5

Find the volume of the solid generated when the region enclosed by $f(x)=\frac{1}{2}x^2$ and the line $g(x)=2x$ is rotated around the x-axis.

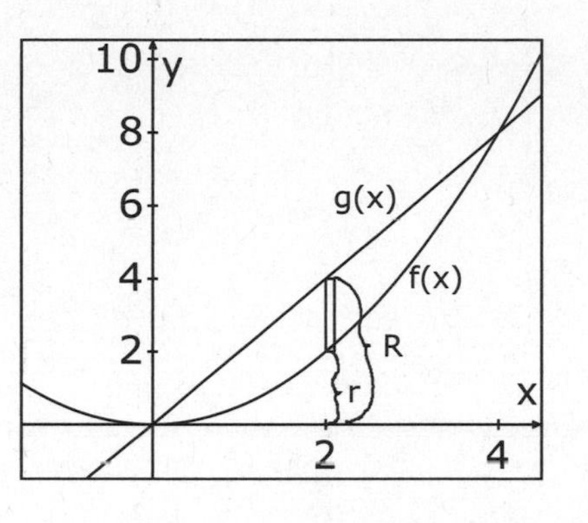

Figure 19-9

The curves intersect where $\frac{1}{2}x^2=2x$.

$x^2-4x=0 \Rightarrow x=0$ or $x=4$

$$V=\int_0^4[\pi\cdot g(x)^2-\pi\cdot f(x)^2]dx$$

$$V=\pi\int_0^4\left[(2x)^2-\left(\frac{1}{2}x^2\right)^2\right]dx$$

$$V=\pi\int_0^4\left[(2x)^2-\left(\frac{1}{2}x^2\right)^2\right]dx=\pi\int_0^4\left(4x^2-\frac{1}{4}x^4\right)dx$$

$$\pi\int_0^4\left(4x^2-\frac{1}{4}x^4\right)dx=\pi\left(\frac{4}{3}x^3-\frac{1}{20}x^5\right)\Bigg|_0^4\approx 34.133\pi$$

Study one more example. This time, try to set it up by yourself before reading through the entire solution.

EXAMPLE 19-6

Let W be the three-sided region in the first quadrant enclosed by $p(x)=e^{(x-2)}$, $g(x)=3-x$, and the y-axis. Sketch the region and one rectangle to follow through the rotation. Find the volume of the solid generated when R is rotated around the x-axis.

Figure 19-10

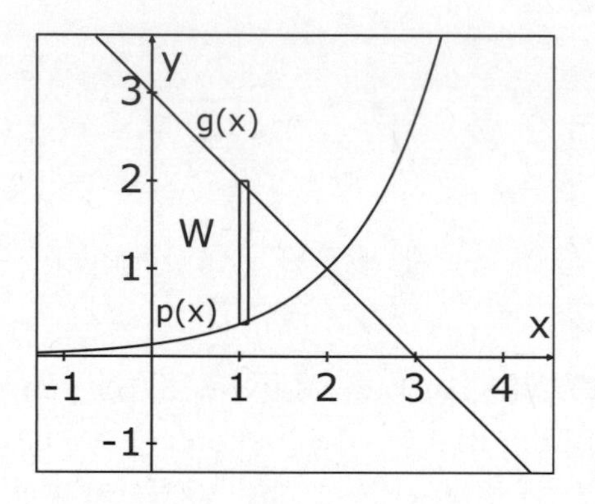

By inspection, both functions contain the point (2,1). The limits of integration are $x=0$ to $x=2$.

The outer radius is defined by the line. The inner radius is defined by the exponential function.

$$\pi\int_0^2 (3-x)^2-\left(e^{(x-2)}\right)^2 dx=\pi\int_0^2 [9-6x+x^2-e^{(2x-4)}]dx$$

$$\pi\int_0^2 [9-6x+x^2-e^{(2x-4)}]dx=\pi\left(9x-3x^2+\frac{1}{3}x^3-\frac{1}{2}e^{(2x-4)}\right)\Bigg|_0^2\approx 8.176\pi$$

Arc Length

Essentially, any quantity that can be added up in infinitely small quantities can be measured with a definite integral. Imagine that there is a curved line on a page and you want to know its length. Unfortunately, all you have in your possession is a ruler. You may not be able to get the exact length, but you can get a pretty good estimate. Do you know how? On very short intervals, all smooth curves have local linearity. Simply take your ruler, measure a number of straight segments along the curve, and add up the lengths as shown in Figure 19-11. You are using the calculus of estimation!

Figure 19-11

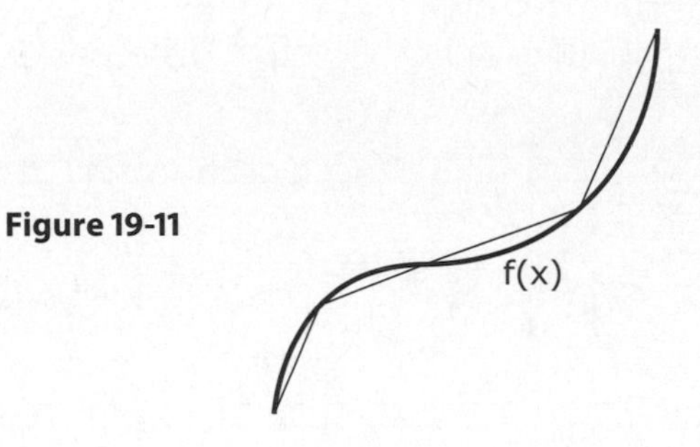

Of course, the more segments you can add, the better your approximation will be. If you are thinking like a calculus master, you already know where this is heading. Let the limit of the number of segments increase infinitely, and the exact arc length will be produced via a definite integral.

Given a differentiable function $y=f(x)$ on the interval $[a,b]$. The length of the arc from $[a,f(a)]$ to $[b,f(b)]$ is $L=\int_a^b \sqrt{1+\left(\frac{dy}{dx}\right)^2}\,dx$.

The concept of adding up a lot of segments is simple, but the algebra behind it takes a bit more thought. Begin by considering a zoomed view of one small arc of a function $y=f(x)$, as shown in Figure 19-12. The segment labeled ds connecting the endpoints of the arc is an estimation of the arc length.

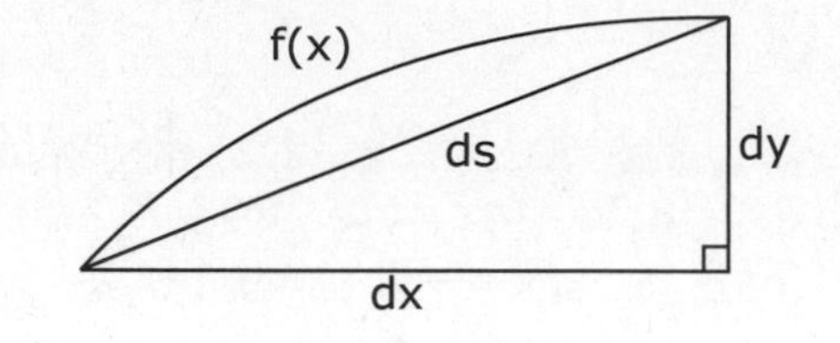

Figure 19-12

Using the Pythagorean Theorem, you find that for the ith arc, $(ds_i)^2 = (\Delta x_i)^2 + (\Delta y_i)^2$, or $ds_i = \sqrt{(\Delta x_i)^2 + (\Delta y_i)^2}$. Factoring out a $(\Delta x_i)^2$ produces a more user-friendly form of the differential, $ds_i = \sqrt{\left[1 + \frac{(\Delta y_i)^2}{(\Delta x_i)^2}\right](\Delta x_i)^2} = \sqrt{1 + \left(\frac{\Delta y_i}{\Delta x_i}\right)^2}\,\Delta x_i$. The length of the arc is the Riemann sum $\lim_{n \to \infty} \sum_{i=1}^{n} \sqrt{1 + \left(\frac{\Delta y_i}{\Delta x_i}\right)^2}\,\Delta x_i = \int_a^b \sqrt{1 + \left(\frac{dy}{dx}\right)^2}\,dx$.

Unless a problem is contrived to produce a very nice radicand, the antidifferentiation often requires a calculator. Example 19-7 illustrates a nicely contrived radicand, and Example 19-8 reviews the process of using the calculator to evaluate the integrand.

EXAMPLE 19-7

Find the length of the curve $y = 2 + \frac{1}{3}x^{(3/2)}$ on the interval $[0,5]$.

$$\frac{dy}{dx} = 0 + \frac{1}{3}\cdot\frac{3}{2}x^{(1/2)} = \frac{1}{2}\sqrt{x} \text{ and } \left(\frac{dy}{dx}\right)^2 = \frac{1}{4}x$$

$$L = \int_0^5 \sqrt{1 + \frac{1}{4}x}\,dx$$

Let $u = 1 + \frac{1}{4}x$ and $du = \frac{1}{4}dx$. The limits become $u = 1$ and $u = 1 + \frac{5}{4} = \frac{9}{4}$.

$$L = \int_1^{9/4} u^{(1/2)}\,4\,du = 4\cdot\frac{2}{3}u^{(3/2)}\Big|_1^{9/4}$$

$$4\cdot\frac{2}{3}u^{(3/2)}\Big|_1^{9/4} = \frac{8}{3}\left(\left(9/4\right)^{(3/2)} - 1^{(3/2)}\right) = \frac{19}{3}$$

EXAMPLE 19-8

Before the vertical spans are attached to the deck, a cable on a suspension bridge hangs in the shape of a catenary curve. The cable is suspended from towers with bases 330 feet apart. The equation to model the cable is $f(x)=\dfrac{e^{(x/20)}+e^{(-x/20)}}{40}$. Find the length needed for one cable.

Imagine the cable on the xy-coordinate system symmetric around the y-axis. Find the length of the curve between $x=-165$ and $x=165$.

$$f'(x)=\frac{\frac{1}{20}e^{(x/20)}-\frac{1}{20}e^{(-x/20)}}{40}=\frac{e^{(x/20)}-e^{(-x/20)}}{800}$$

$$[f'(x)]^2=\left(\frac{e^{(x/20)}-e^{(-x/20)}}{800}\right)^2=\frac{e^{(x/10)}-2+e^{(-x/10)}}{640000}$$

$$L=\int_{-165}^{165}\sqrt{1+\frac{e^{(x/10)}-2+e^{(-x/10)}}{640000}}\,dx\approx 442\text{ ft.}$$

Skill Check

The following Skill Check offers one more set of questions to help you practice applying the ideas introduced in this section. They are by no means meant to be comprehensive, but they will give you an opportunity to test your abilities. As always, try to do as much of the work as possible without a calculator, relying on tools you have acquired as you have moved through this book. Sketching regions, finding points of intersection, and finding antiderivatives by hand are all skills in which you have significantly improved since the beginning of this book. As always, many more resources and practice problems are available on the Internet.

1. Find the area between the graphs of $f(x)=\dfrac{5x}{1+x^2}$ and $g(x)=\dfrac{1}{2}x^2$. Use a graphing calculator to find the points of intersection.

2. Find the area of region Q enclosed by the graphs of $p(x)=x^{(2/3)}$, $h(x)=-\sqrt{x}$, and $r(x)=2-x$. The region is shown in Figure 19-13.

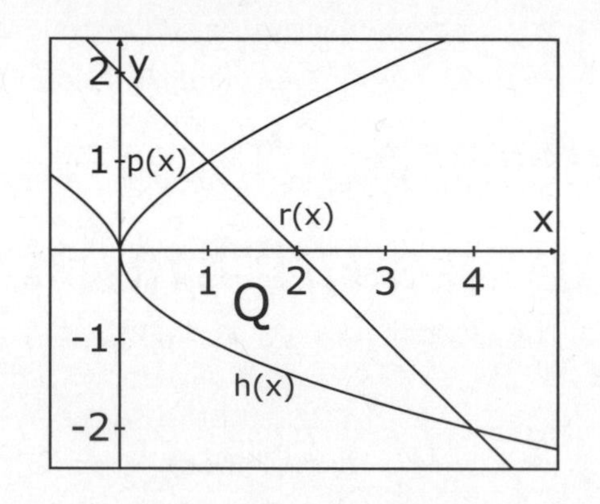

Figure 19-13

3. The base of a solid lies in the xy-plane and has cross sections perpendicular to the x-axis that are equilateral triangles. One side of each triangle runs from the x-axis to the function $g(x)=3-x$ on the interval [0,3]. Find the volume of the solid.
4. The region below $h(x)=\dfrac{2\ln(x)}{\sqrt{x}}$ and above the x-axis on the interval $[1,e]$ is rotated around the x-axis. Find the volume of the solid generated.
5. The region enclosed by the graph of $k(x)=x^2+1$ and $f(x)=x+3$ is rotated around the x-axis. Find the volume of the solid generated.
6. Set up an integral to find the length of the arc of $y=\sin(2x)$ on the interval [0,5]. Use your calculator to evaluate.

APPENDIX A

Useful Prerequisite Information

This appendix summarizes many of the formulas and graphs discussed in Chapter 1 on prerequisite skills. It contains key ideas, equations, and formulas from algebra, geometry, and trigonometry that are used regularly in calculus. The information is not exhaustive. Numerous opportunities to study any of these topics and to practice these skills are available on the Internet. You can find them by using any search engine.

Algebra

The key skills from algebra that are important in calculus include, but are not limited to, writing equations of lines, working with exponents and roots, and graphing linear and quadratic functions. Factoring is also an important skill, which is discussed in more detail in Chapter 1.

Slope of a Line Given Two Points

The slope between two points (x_1, y_1) and (x_2, y_2) is $m = \dfrac{y_2 - y_1}{x_2 - x_1}$.

Forms of Linear Equations

Slope-intercept form of a line: $y = mx + b$, where m is the slope and b is the y-intercept

Point-slope form of a line: $y - y_1 = m(x - x_1)$, where (x_1, y_1) is any known point on the line

Equation of a horizontal line through a point (a,b): $y = b$

Equation of a vertical line through a point (a,b): $x = a$

Quadratic Equations

Standard form of a quadratic function: $y = ax^2 + bx + c$

x-coordinate of the vertex: $x = \dfrac{-b}{2a}$

y-intercept: $y = c$

Vertex form of a quadratic function: $y = a(x - h)^2 + k$, where (h,k) is the vertex

Direction of the graph: $a > 0$ causes parabola to open upward. $a < 0$ causes parabola to open downward.

Factoring Patterns

Perfect square trinomial: $a^2 + 2ab + b^2 = (a + b)^2$

Difference of two squares: $a^2 - b^2 = (a + b)(a - b)$

Difference of two cubes: $a^3 - b^3 = (a - b)(a^2 + ab + b^2)$

Sum of two cubes: $a^3 + b^3 = (a + b)(a^2 - ab + b^2)$

Properties of Exponents

Multiplying powers causes the exponents to add: $x^a \cdot x^b = x^{a+b}$

Dividing powers causes the exponents to subtract: $\frac{x^a}{x^b} = x^{a-b}$

Raising a power to an exponent multiplies the exponents: $\left(x^a\right)^b = x^{(a \cdot b)}$

Negative exponents reciprocate and change sign: $x^{-a} = \frac{1}{x^a}$

Fractional powers are roots: $x^{(1/a)} = \sqrt[a]{x}$

Geometry

There is a large amount of geometry material in calculus. Being able to visualize calculus concepts is a big step toward gaining a deeper understanding of the subject. Geometry topics that are frequently in calculus include areas of various shapes, volumes of solids, Pythagorean relationships in right triangles, and the distance between two points. Many of the algebra concepts also resurface in the context of geometry.

Area Formulas

Rectangle: $A = L \cdot W$, where L is the length and W is the width

Triangle: $A = \frac{1}{2} b \cdot h$, where b is the base and h is the height measured perpendicular to the base

Equilateral Triangle: $A = \frac{s^2}{4} \cdot \sqrt{3}$, where s is the length of a side

Circle: $A = \pi \cdot r^2$, where r is the radius—that is, the distance from the center to the circle

Trapezoid: $A = \frac{1}{2} \cdot h \cdot (b_1 + b_2)$, where b_1 and b_2 are the bases—that is, the parallel sides—and h is the perpendicular distance between the bases

Volume Formulas

Cylinder: $V = \pi \cdot r^2 \cdot h$, where r is the radius of the base, and h is the height of the cylinder

Cone: $V = \frac{1}{3}\pi \cdot r^2 \cdot h$, where r is the radius of the base, and h is the perpendicular height of the cone

Washer: $V = \pi \cdot R^2 \cdot h - \pi \cdot r^2 \cdot h$, where R is the outer radius, r is the inner radius, and h is the height

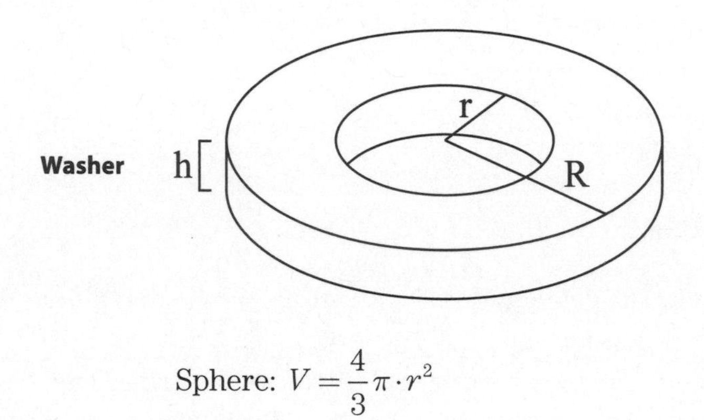

Sphere: $V = \frac{4}{3}\pi \cdot r^2$

Pythagorean Theorem

The sum of the squares of the legs of a right triangle is equal to the square of the hypotenuse. If a and b are the lengths of the legs, and c is the length of the hypotenuse, then $a^2 + b^2 = c^2$.

Distance Between Two Points

If (x_1, y_1) and (x_2, y_2) are two points in the coordinate plane, the straight-line distance D between the points is $D = \sqrt{(x_2 - x_1)^2 + (y_2 - y_1)^2}$.

Trigonometry

The primary trigonometry skills necessary for calculus are right triangle trigonometry, unit circle values, and select trigonometric identities. The ability to quickly sketch the sine and cosine functions and slight transformations of those functions is also imperative.

Trigonometric Ratios in Right Triangles

The six trigonometric ratios in a right triangle are sine, cosine, tangent, secant, cosecant, and cotangent.

For any acute angle in a right triangle, call it $\angle P$,

$$\sin(\angle P) = \frac{\text{Opposite side}}{\text{Hypotenuse}} \quad \cos(\angle P) = \frac{\text{Adjacent side}}{\text{Hypotenuse}} \quad \tan(\angle P) = \frac{\text{Opposite side}}{\text{Adjacent side}}$$

$$\csc(\angle P) = \frac{\text{Hypotenuse}}{\text{Opposite side}} \quad \sec(\angle P) = \frac{\text{Hypotenuse}}{\text{Adjacent side}} \quad \cot(\angle P) = \frac{\text{Adjacent side}}{\text{Opposite side}}$$

Trigonometric Identities

The reciprocal identities:

$$\csc(x) = \frac{1}{\sin(x)} \qquad \sec(x) = \frac{1}{\cos(x)} \qquad \cot(x) = \frac{1}{\tan(x)}$$

The Pythagorean identities:

$$\sin^2(x) + \cos^2(x) = 1 \qquad \tan^2(x) + 1 = \sec^2(x) \qquad \cot^2(x) + 1 = \csc^2(x)$$

The double-angle identities:

$$\sin(2x) = 2\cos(x)\sin(x) \qquad \cos(2x) = \cos^2(x) - \sin^2(x)$$

The power-reducing identities:

$$\cos^2(x) = \frac{1+\cos(2x)}{2} \qquad \sin^2(x) = \frac{1-\cos(2x)}{2}$$

Nine Graphs Common to Calculus

The nine graphs that are the foundation of the majority of graphs in calculus are shown here. You should be able to quickly sketch any of these graphs and simple transformations of these graphs. You should also know the domain and range of each of these functions and be able to determine how the domain and range are changed by transformations to the functions.

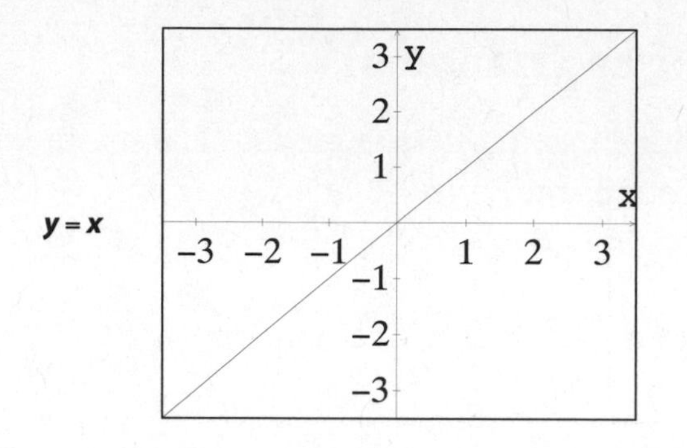

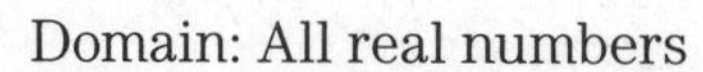
Domain: All real numbers

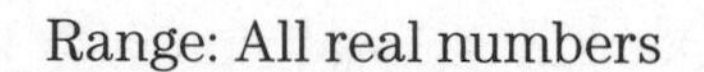
Range: All real numbers

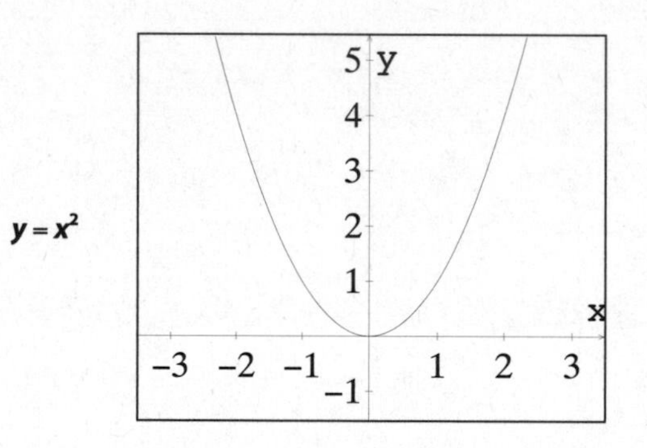

Domain: All real numbers   Range: $y \geq 0$

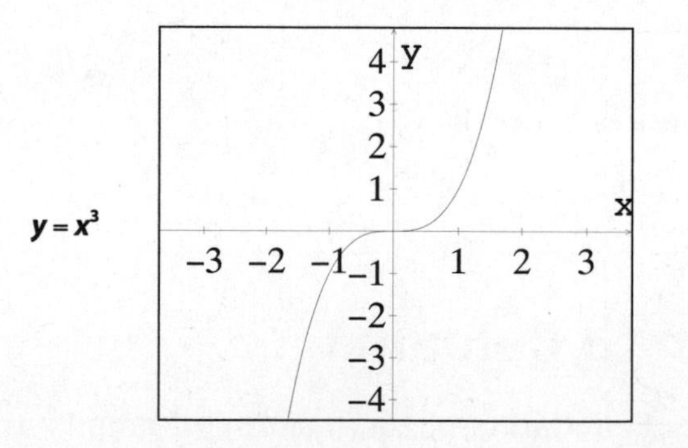

Domain: All real numbers   Range: All real numbers

$y = |x|$

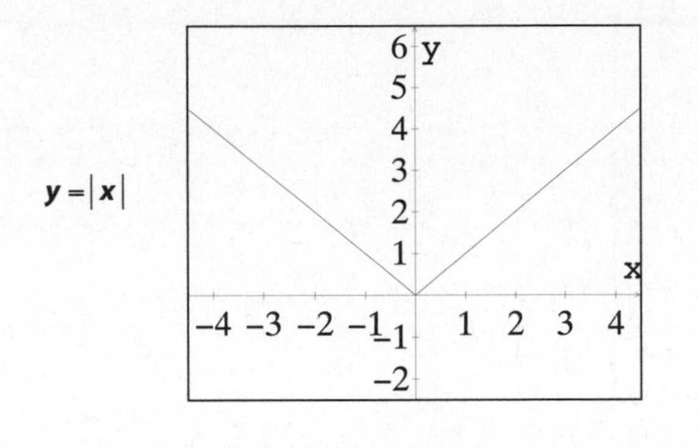

Domain: All real numbers   Range: $y \geq 0$

$y = \sin(x)$

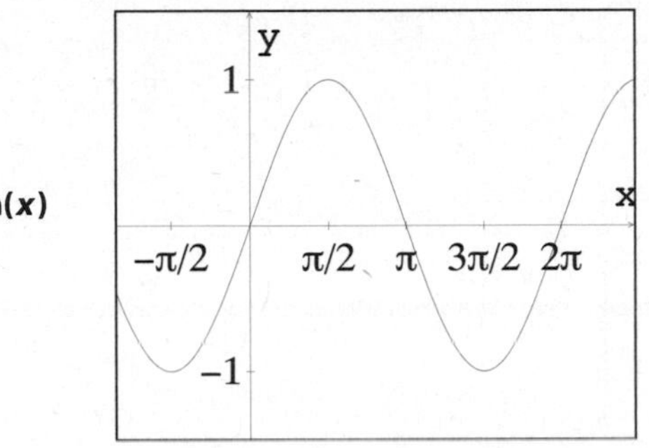

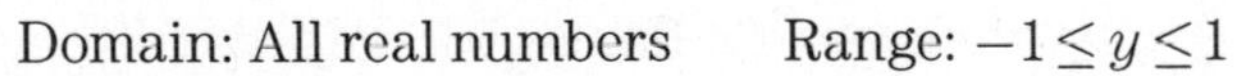
Domain: All real numbers   Range: $-1 \leq y \leq 1$

$y = \cos(x)$

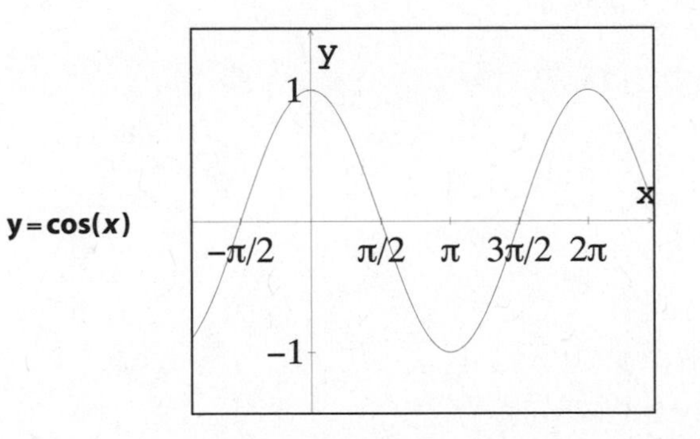

Domain: All real numbers   Range: $-1 \leq y \leq 1$

$y = e^x$

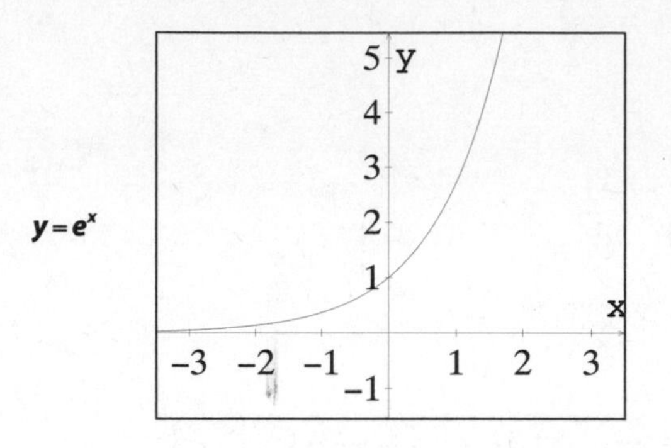

Domain: All real numbers Range: $y > 0$

$y = Ln(x)$

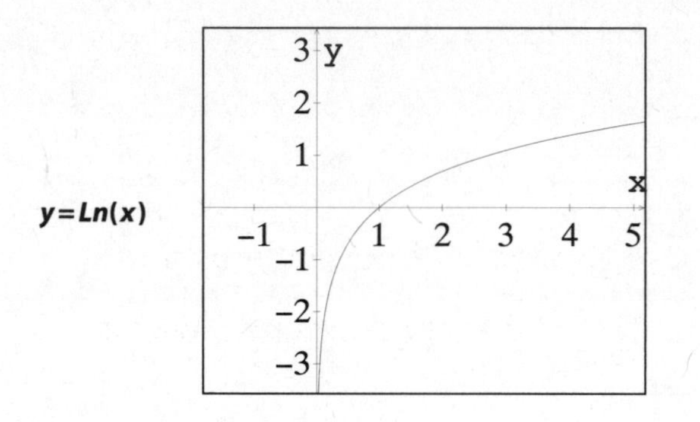

Domain: $x > 0$

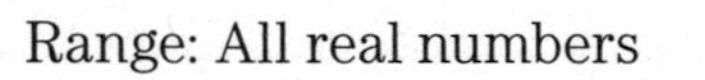
Range: All real numbers

$y = \frac{1}{x}$

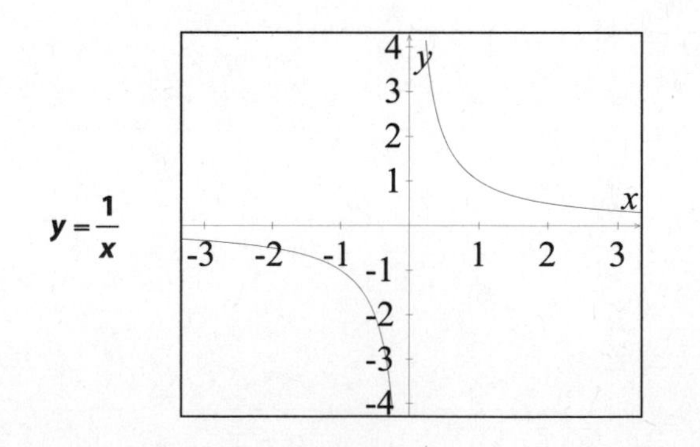

Domain: $x < 0 \cup x > 0$ Range: $y < 0 \cup y > 0$

APPENDIX B

Derivatives and Integrals

This appendix contains a summary of the primary derivatives and integrals that must be committed to memory. The formulas are the building blocks of much of the work done in any beginning calculus course. Use the tables frequently as a reference until you are confident of the facts contained within each table.

Derivative and Integral Rules

As you examine the accompanying tables, note that there are strong similarities between the two. That is because an indefinite integral essentially reverses the process of differentiation. If you have just begun to read this book, these formulas may not mean anything to you yet, but in time they will. As you need to refer to them, consult this section frequently. Over time, the facts set out here will become second nature to you, and your reliance on the tables will diminish. Be aware that there are hundreds more integral formulas than are presented here, but most of those are meant to handle some very obscure or unique integration situations.

Throughout the tables, u and v are functions of an independent variable, and all other variables are assumed to be constants. Nearly all of the derivative formulas are written in differential form to allow for the independent variable to be any variable.

THE MOST COMMON DERIVATIVE FORMULAS

	Function	Derivative
1.	$y = u^n$ (n is any real number)	$dy = n \cdot u^{(n-1)} \cdot du$
2.	$y = k$ (k is a constant)	$dy = 0$
3.	$y = u \cdot v$	$dy = u \cdot dv + v \cdot du$
4.	$y = \frac{u}{v}$	$dy = \frac{v \cdot du - u \cdot dv}{v^2}$
5.	$y = \sin(u)$	$dy = \cos(u) \cdot du$
6.	$y = \cos(u)$	$dy = -\sin(u) \cdot du$
7.	$y = \tan(u)$	$dy = \sec^2(u) \cdot du$
8.	$y = \cot(u)$	$dy = -\csc^2(u) \cdot du$
9.	$y = \sec(u)$	$dy = \sec(u) \cdot \tan(u) \cdot du$
10.	$y = \csc(u)$	$dy = -\csc(u) \cdot \cot(u) \cdot du$
11.	$y = \sin^{-1}(u)$	$dy = \frac{du}{\sqrt{1-u^2}}$
12.	$y = \cos^{-1}(u)$	$dy = \frac{-du}{\sqrt{1-u^2}}$

	Function	Derivative
13.	$y = \tan^{-1}(u)$	$dy = \frac{du}{1+u^2}$
14.	$y = \cot^{-1}(u)$	$dy = \frac{-du}{1+u^2}$
15.	$y = \sec^{-1}(u)$	$dy = \frac{du}{\lvert u \rvert \sqrt{u^2-1}}$
16.	$y = \csc^{-1}(u)$	$dy = \frac{-du}{\lvert u \rvert \sqrt{u^2-1}}$
17.	$y = e^u$	$dy = e^u \cdot du$
18.	$y = a^u$	$dy = a^u \cdot \ln(a) \cdot du$
19.	$y = \ln(u)$	$dy = \frac{du}{u}$
20.	$y = \log_a u$	$dy = \frac{1}{\ln(a)} \cdot \frac{du}{u}$

Derivative Rules for Special Cases

Chain rule: If $y = f(g(x))$, then $\frac{dy}{dx} = f'(g(x)) \cdot g'(x)$.

Function raised to a function: If $y = u^v$, then use the logarithmic differentiation method.

THE MOST COMMON INTEGRAL FORMULAS

	Integral	Antiderivative
1.	$\int u^n \cdot du$	$\frac{u^{(n+1)}}{n+1} + C$
2.	$\int k \cdot du$ (k is a constant)	$k \cdot u + C$
3.	$\int \cos(u) \cdot du$	$\sin(u) + C$
4.	$\int \sin(u) \cdot du$	$-\cos(u) + C$
5.	$\int \sec^2(u) \cdot du$	$\tan(u) + C$

THE MOST COMMON INTEGRAL FORMULAS

	Integral	Antiderivative
6.	$\int \csc^2(u)\cdot du$	$-\cot(u)+C$
7.	$\int \sec(u)\cdot \tan(u)\cdot du$	$\sec(u)+C$
8.	$\int \csc(u)\cdot \cot(u)\cdot du$	$-\csc(u)+C$
9.	$\int \frac{du}{\sqrt{1-u^2}}$	$\sin^{-1}(u)+C$
10.	$\int \frac{-du}{\sqrt{1-u^2}}$	$\cos^{-1}(u)+C$
11.	$\int \frac{du}{1+u^2}$	$\tan^{-1}(u)+C$
12.	$\int \frac{-du}{1+u^2}$	$\cot^{-1}(u)+C$
13.	$\int \frac{du}{u\sqrt{u^2-1}}$, $u \geq 1$	$\sec^{-1}(u)+C$
14.	$\int \frac{-du}{u\sqrt{u^2-1}}$, $u \geq 1$	$\csc^{-1}(u)+C$
15.	$\int e^u \cdot du$	$e^u + C$
16.	$\int a^u du$	$\frac{a^u}{ln(a)}+C$
17.	$\int \frac{du}{u}$	$ln\lvert u \rvert + C$

APPENDIX C

The Final Exam

Of course, this is not really a final exam. What follows is simply a way for you to measure your progress toward a personal goal of increasing knowledge about the subject of calculus. Reading through an informational book or practicing a few problems per chapter by no means guarantees mastery. The only way to get better and better at calculus is to continue to be a student of the subject. Enjoy it. Practice it. Read about it. And talk about it with others who share your interest in calculus.

Final Exam Time

The questions that follow are intended to provide a broad cross section of the material in a typical first-year calculus course. Hundreds of questions could be asked, over a wide range of topics, but fifty or so is a good start. Feel free either to try them all after reading the entire book or to tackle questions piecemeal after reading each chapter. Remember, the Internet is a great source of additional practice problems. Be careful, though, because the scope of some the questions you find on the Internet will be well beyond what has been presented here as first-year calculus material.

This "final exam" begins with a series of true-or-false questions intended to test larger conceptual understanding. Solutions to the remaining questions should be worked out by hand. Most questions are intended to be done without a calculator, but the intention is growth in knowledge, not a graded assessment, so feel free to use a calculator as needed. When it is explicitly intended that you use a calculator, the instructions will say so.

True-or-False Questions

1. An inflection point on a function $g(x)$ always occurs where $g''(x)=0$.

2. If $f'(x)$ exists at a point $x=a$, then function f is continuous at a.

3. If $f(x)$ is continuous at $x=a$, then $f'(x)$ must exist.

4. If $[b,h(b)]$ is a point where the function $h(x)$ has a local maximum, then $h'(b)=0$ and $h''(b)<0$.

5. $\int \frac{du}{g(u)} = \ln|g(u)|+C$

6. The average rate of change of a function $f(x)$ on the interval $[a,b]$ is calculated by $\frac{f(a)+f(b)}{2}$.

7. The total area between a function $p(x)$ and the x-axis on an interval $[a,b]$ is found by $\int_a^b p(x)dx$.

8. The definite integral of velocity over a given interval calculates displacement.

9. If $k(x)$ is continuous for all real numbers, and $k(1)=7$ and $k(5)=-4$, then $k(x)=0$ has at least one solution in the interval $[1,5]$.

10. Calculus is essentially the study of change.

Problems

The following problems are intended to be worked out by hand, with occasional use of a calculator.

11. Evaluate the limit. $\lim_{x\to 5}\frac{4x^2-20x}{3x-15}$

12. Evaluate the limit. $\lim_{x\to -\infty}\frac{7x^3+4\cos(x)}{\left|2x^3\right|}$

13. Evaluate the limit. $\lim_{x\to 4^-} f(x)$ given that $f(x)=\begin{cases}\sin\left(\frac{\pi}{x}\right) & \text{for } x<4\\ 3x+2 & \text{for } x>4\end{cases}$

14. Determine whether or not the function $r(x)$ is continuous at $x=-2$ given that $r(x)=\begin{cases}\sqrt{7-x} & \text{for } x<-2\\ x^2-x-3 & \text{for } x\geq -2\end{cases}$.

15. Find the average rate of change of the function $s(x)=x\sqrt{2x+1}$ on the interval $[4,12]$.

16. Use the definition of a derivative to evaluate $\lim_{h\to 0}\frac{2^{(3+h)}-8}{h}$ without a calculator.

17. Given $f(x)=\begin{cases}ax^2-bx & \text{for } x>2\\ ax^3+b+1 & \text{for } x\leq 2\end{cases}$, find a and b such that $f'(2)$ exists.

18. The line tangent to the graph of $y=\sin(x)$ at $x=\frac{\pi}{3}$ intersects the graph in one other point. Write the equation of the tangent line by hand, and then use a graphing calculator to find the point of intersection.

19. If $f(x) = \csc(3x)$, find $f'(x)$.

20. If $g(x) = \dfrac{\sin(x)}{2 + \cos(x)}$, find $g'\left(\dfrac{\pi}{2}\right)$.

21. Find $\dfrac{dy}{dx}$ if $y = \sin(u^2)$ and $u = \tan(3x)$. The final answer should be in terms of just x.

22. Find the slope of the line normal to graph of the curve $x^2 + \sin(xy) + y^3 = 8$ at the y-intercept.

23. Find the instantaneous rate of change of $p(t) = \sqrt{4t+1} \cdot e^{3t}$ when $t = 0$.

24. Find the x-coordinate where the graph of $y = \ln(5x^2 - 3x + 1)$ has a horizontal tangent.

25. Find $\dfrac{d^2y}{dx^2}$ if $y = 5^{(x^2)}$.

26. Given the $f(x) = x^{\tan(x)}$, find $f'(x)$.

27. Find the slope of the tangent to $g(x) = \cos^{-1}(x^2)$, where $g(x) = \dfrac{\pi}{3}$ and $x > 0$.

28. Find the derivative of $y = x^2 \cdot \csc^{-1}(x)$.

29. Sketch the derivative of the graph of $p(x)$ shown the following figure.

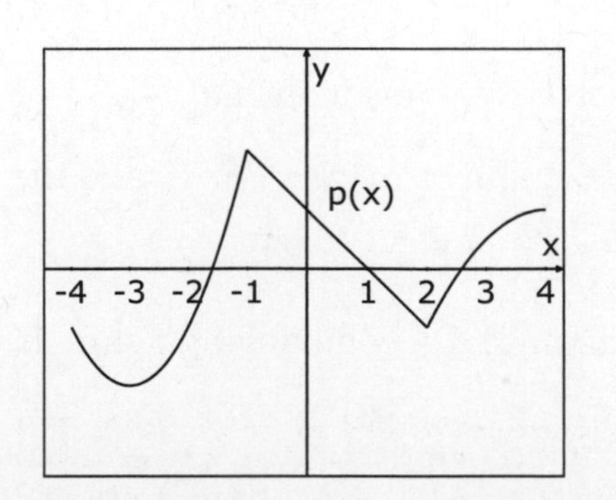

30. Find the x-coordinate of each local extremum on $f(x)=x^3+5x^2-8x+1$. Use the second-derivative test for local extremes to justify your conclusion.

31. Find the x-coordinates of all points of inflection on the function $h(x)=\frac{x^2+1}{e^x}$.

32. A right circular cylinder with no top has a fixed surface area of 16 square inches. Find the exact radius that produces a can with maximum volume.

33. A point is moving away from the origin along the graph of $y=x^2$. When the point is at $(2,4)$, its straight-line distance from the origin is increasing at a rate of 9 units per second. At what rate is its y-coordinate changing?

34. Find the c value that satisfies the Mean Value Theorem for the function $s(x)=2x^3-1$ on the interval $[0,2]$.

35. Find $LRAM_6$ for $g(x)=3+\cos(2x)$ on the interval $[0,\pi]$.

36. Estimate the value of the definite integral $\int_1^6 \sqrt{x+1}\,dx$ using the trapezoidal rule with five trapezoids.

37. Find the actual value of $\int_1^6 \sqrt{x+1}\,dx$ by antidifferentiation. Then use a calculator to find the difference between the actual result and the trapezoidal approximation.

38. Find the area between the graph of $h(x)=2^x-4$ and the x-axis on the interval $[0,3]$.

39. The graph of $f(t)$ is shown in the accompanying figure. If $a(x)=\int_1^x f(t)\,dt$, find the x-coordinate of each inflection point on $a(x)$.

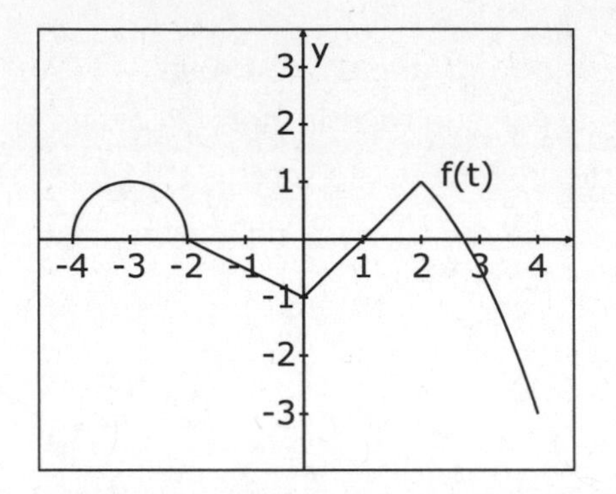

40. Find the instantaneous rate of change of $k(x)=\int_{2}^{x^2}\frac{\sqrt[3]{t-1}}{t+2}\,dt$ when $x=3$.

41. Integrate using u-substitution. $\int \frac{3x+6}{x^2+4x+5}\,dx$.

42. Which differential equation is represented by the graph of the slope-field in the accompanying figure? Explain why the option chosen is correct and the options rejected are not correct.

A. $\frac{dy}{dx}=\frac{1}{\ln|x|}$ B. $\frac{dy}{dx}=x^{\left(-\frac{2}{3}\right)}$ C. $\frac{dy}{dx}=x^{\left(-\frac{1}{3}\right)}$ D. $\frac{dy}{dx}=\frac{1}{x^2+1}$

43. Find the particular solution to $\frac{dy}{dx}=4x\sqrt{y}$ if $y(1)=9$.

44. Find the average value of the function $y=\frac{4}{2x+1}$ on the interval $[1,7]$.

45. A particle is moving along a horizontal axis. Its velocity in feet per minute is given by $v(t)=t^{(2/3)}-1$. Find the total distance that it traveled in the first 8 minutes.

46. Let R be the first-quadrant region bounded by $f(x)=5-\frac{1}{2}x^2$, $g(x)=1+\frac{1}{4}x^3$, and the y-axis. Find the area of region R.

47. Let P be the region below $m(x)=\sqrt{\sec\left(\frac{x}{3}\right)\tan\left(\frac{x}{3}\right)}$ and above the x-axis on the interval $[0,\pi]$. Find the volume of the solid generated when P is rotated around the x-axis.

48. Set up but do not integrate an integral expression for the volume of the solid generated when the region bounded by $a(x)=x^x$ and $b(x)=3x-2$ is rotated around the x-axis. Find the limits of integration by graphing the functions.

49. Find the volume of a solid whose cross sections are squares perpendicular to the x-axis between $x=0$ and $x=2$. The diagonal of each square runs from the x-axis to the curve $y=e^{(-x^2)}\cdot\sqrt{x}$.

50. Find the length of the curve $c(x)=\sqrt{x^2+1}$ on the interval [–1,3]. Set up the integral by hand, and then evaluate it with a calculator.

APPENDIX D

Answer Key

Chapter 2

1. $\lim_{x \to -1} w(x) = 2$

The y values are approaching 2 from each side of $x = -1$, even though $w(-1) = -2$.

2. $\lim_{x \to 1^-} w(x) = 2$

Remember that the minus sign means approach 1 from the left. Examine y values when x is just less than 1. The y values are headed toward 2.

3. $\lim_{x \to 1^+} w(x) = -1$

For values of x just slightly larger than 1, the graph is going to $y = -1$.

4. $\lim_{x \to 1} w(x)$ does not exist.

The answers to Problems 2 and 3 show that the left- and right-sided limits are not equal.

5. $\lim_{x \to 3^-} w(x) = 3$

For values of x just slightly less than 3, the graph is going toward $y = 3$. The limit exists, even though $w(3)$ does not exist.

6. $\lim_{x \to 4} \dfrac{x^2 - 5x + 4}{x - 4} = 3$

$$\begin{aligned}\lim_{x \to 4} \frac{x^2 - 5x + 4}{x - 4} &= \lim_{x \to 4} \frac{(x-4)(x-1)}{x-4} \\ &= \lim_{x \to 4} (x - 1) \\ &= 3\end{aligned}$$

7. $\lim_{x \to \infty} \dfrac{7 - 5x^3}{8x^3 + 2x^2} = -\dfrac{5}{8}$

$$\begin{aligned}\lim_{x \to \infty} \frac{7 - 5x^3}{8x^3 + 2x^2} \cdot \frac{1/x^3}{1/x^3} &= \lim_{x \to \infty} \frac{\frac{7}{x^3} - 5}{8 + \frac{2}{x}} \\ &= \frac{0 - 5}{8 + 0} \\ &= -\frac{5}{8}\end{aligned}$$

8. $\lim_{x \to -4} \sqrt{12 - x} = 4$

Use direct substitution.

$$\begin{aligned}\lim_{x \to -4} \sqrt{12 - x} &= \lim_{x \to -4} \sqrt{12 - (-4)} \\ &= \sqrt{16} \\ &= 4\end{aligned}$$

9. $\lim_{x \to -\infty} \cos\left(\dfrac{1}{x}\right) = 1$

Use intuitive thinking. As x approaches $-\infty$, then $\dfrac{1}{x}$ approaches 0, but all values will be negative. That does not influence the answer, because $\cos(0) = 1$.

Let $a = \dfrac{1}{x}$. Then $\lim_{x \to -\infty} \cos\left(\dfrac{1}{x}\right) = \lim_{a \to 0^-} \cos(a)$

$$\begin{aligned}&= \cos(0) \\ &= 1\end{aligned}$$

10. $\lim_{x \to 7^+} \frac{x+1}{x-7} \Rightarrow \infty$ Many books also say, "Does not exist" because the function grows without bound. Either is acceptable, depending on the convention adopted by the instructor or by textbook being used.

Use numerical thinking. As x gets very close to 7 from the right, the numerator is approaching 8, but the denominator is a positive value getting closer and closer to 0. The ratio of these values gets larger and larger.

Chapter 3

1. Factor and reduce.

$$h(x) = \frac{2x-6}{x^2+4x-21} = \frac{2(x-3)}{(x-3)(x+7)} = \frac{2}{x+7}$$

Because $(x-3)$ canceled, there is a removable discontinuity at its zero, $x=3$. Because $(x+7)$ remained in the denominator, there is a vertical asymptote (or infinite discontinuity) at $x=-7$.

2. Factor and reduce to find the removable discontinuity. The numerator is a difference of two cubes.

$$f(x) = \frac{x^3-8}{x-2} = \frac{(x-2)(x^2+2x+4)}{x-2} = x^2+2x+4$$

When $x=2$, the y value is found by substituting into the reduced form of the function.

$$y = 2^2 + 2 \cdot 2 + 4 = 12$$

$$f(x) = \begin{cases} \frac{x^3-8}{x-2} & \text{for} \quad x \neq 2 \\ 12 & \text{for} \quad x = 2 \end{cases}$$

3. At $x=2$, the left- and right-sided limits must be equal.

$$\lim_{x \to 2^-} (ax^2 - 3) = \lim_{x \to 2^+} (5x+1)$$

$$a \cdot 2^2 - 3 = 5 \cdot 2 + 1$$

$$4a - 3 = 11$$

$$a = \frac{7}{2}$$

4. There is a jump discontinuity at $x=-4$ and a removable discontinuity at $x=-2$. The graph is continuous at $(2,1)$.

Intervals of continuity are $[-6,-4] \cup (-4,-2) \cup (-2,\infty)$.

5. At each junction of the domain, check the appropriate limits.

At $x=-3$, $\lim_{x \to -3^-} (-x+2) = 5$ and $\lim_{x \to -3^+} x^2 + x = (-3)^2 + (-3) = 6$. Because the limits are not equal, there is a discontinuity at $x=-3$.

At $x=1$, $\lim_{x \to 1^-} x^2 + x = (1)^2 + (1) = 2$ and $\lim_{x \to 1^+} \sqrt{x+3} = \sqrt{4} = 2$. Because the limits are equal and $q(1)$ is defined, $q(x)$ is continuous at $x=1$.

6. Solving $x^4 - 2x = 3$ is the same as solving $x^4 - 2x - 3 = 0$.

   Let $h(x) = x^4 - 2x - 3$.

   $h(1) = 1^4 - 2 \cdot 1 - 3 = -4$ and
   $h(2) = 2^4 - 2 \cdot 2 - 3 = 9$.

   Because all polynomials are continuous for all real numbers, $h(1)$ is negative, and $h(2)$ is positive, the Intermediate Value Theorem assures us that $h(x) = 0$ must have at least one solution between $x = 1$ and $x = 2$.

Chapter 4

1. The average rate of change is the change in gallons divided by the change in time. The gallons drop from 42,500 to 34,000. It is 4 hours from 10 A.M. 2 P.M.

   $$\frac{\Delta g}{\Delta t} = \frac{34{,}000 - 42{,}500}{4} = -2125 \,{}^{\text{gal}}\!/_{\text{hr}}$$

2. The average rate of change is simply the slope between the points.

   $$R_{\text{avg}} = \frac{p(6) - p(2)}{6 - 2} = \frac{\frac{1}{6} - \frac{1}{2}}{4} = -\frac{1}{12}$$

3. Without a function, the best estimate is found by using the shortest available time interval containing the moment in question. Calculate the average rate of change of the time interval from 4 to 6 seconds.

   $$\text{Acceleration} \approx \frac{62 - 47 \text{ feet/second}}{6 - 4 \text{ seconds}} \approx 7.5 \text{ feet per second}^2$$

4. How fast an object is moving at a moment in time is its instantaneous rate of change.

   Use $[2, d(2)]$ and $[2 + h, d(2 + h)]$.

   $$\lim_{h \to 0} \frac{d(2+h) - d(2)}{(2+h) - 2} = \lim_{h \to 0} \frac{4.9(2+h)^2 - 4.9 \cdot 2^2}{h}$$
   $$= \lim_{h \to 0} \frac{4.9[(4 + 2h + h^2) - 4]}{h}$$
   $$= \lim_{h \to 0} \frac{4.9h(2+h)}{h}$$
   $$= \lim_{h \to 0} 4.9(2+h)$$
   $$= 9.8 \,{}^{\text{m}}\!/_{\text{sec}}$$

5. Find the point and find the slope at $x = -1$.

   $f(-1) = (-1)^3 - 4 = -5$

   $$\lim_{h \to 0} \frac{f(-1+h) - f(-1)}{h} = \lim_{h \to 0} \frac{[(-1+h)^3 - 4] - (-5)}{h}$$
   $$= \lim_{h \to 0} \frac{-1 + 3h - 3h^2 + h^3 - 4 + 5}{h}$$
   $$= \lim_{h \to 0} \frac{3h - 3h^2 + h^3}{h}$$
   $$= \lim_{h \to 0} \frac{h(3 - 3h + h^2)}{h}$$
   $$= 3$$

   $y - (-5) = 3[x - (-1)]$, or $y = 3x - 2$

6. The derivative of $y=\frac{3}{x-7}$ does not exist at $x=7$ because the graph has a vertical asymptote at $x=7$. The graph has an infinite discontinuity, and a derivative cannot exist at a discontinuity.

7. The derivative $y=|x+3|$ does not exist at $x=-3$ because the graph has a corner there. Transforming the basic graph $y=|x|$ into $y=|x+3|$ shifts it 3 units left. The corner at $(0,0)$ on $y=|x|$ is now at $(-3,0)$ on $y=|x+3|$.

Chapter 5

1. $\lim_{x\to4}\frac{\frac{1}{2}x^2-\frac{1}{2}(4^2)}{x-4}=\frac{1}{2}\lim_{x\to4}\frac{x^2-16}{x-4}$

Set up the limit and factor a $\frac{1}{2}$ out in front.

$\frac{1}{2}\lim_{x\to4}\frac{x^2-16}{x-4}=\frac{1}{2}\lim_{x\to4}\frac{(x-4)(x+4)}{x-4}$ Factor the numerator.

$\frac{1}{2}\lim_{x\to4}\frac{(x-4)(x+4)}{x-4}=\frac{1}{2}\lim_{x\to4}(x+4)$ Cancel $(x-4)$.

$\frac{1}{2}\lim_{x\to4}(x+4)=\frac{1}{2}(4+4)=4$ Evaluate the limit at $x=4$.

2. Comparing $\lim_{x\to7}\frac{\left(\frac{2}{x}\right)-\left(\frac{2}{7}\right)}{x-7}$ to $f'(a)=\lim_{x\to a}\frac{f(x)-f(a)}{x-a}$, the given expression finds the slope of $f(x)=\frac{2}{x}$ at $x=7$ or $(7,\frac{2}{7})$.

3. When $x=1$, $y=1^3-3(1)+5=3$. The ordered pair is $(1,3)$.

The slope is found from the derivative of $y=x^3-3x+5$.

$y'=3x^2-3$ evaluated at $x=1$ gives $y'=3(1)^2-3=0$.

A slope of zero means that the tangent line is a horizontal line through (1,3) with the equation $y = 3$.

4. Differentiate each factor in $h(x) = (x^2 + 9x - 11)(5x^3 - 7x^2 + 13)$ and pair it with the other factor.

$$h'(x) = \frac{d}{dx}(x^2 + 9x - 11) \cdot (5x^3 - 7x^2 + 13) + \frac{d}{dx}(5x^3 - 7x^2 + 13) \cdot (x^2 + 9x - 11)$$

$$h'(x) = (2x + 9) \cdot (5x^3 - 7x^2 + 13) + (15x^2 - 14x) \cdot (x^2 + 9x - 11)$$

5. $g(x) = \dfrac{12x^3 - 8x}{2x}$

$$g(x) = \frac{12x^3}{2x} - \frac{8x}{2x} = 6x^2 - 4$$

Distribute the denominator term and reduce each fraction.

$\dfrac{dg}{dx} = 2 \cdot 6x^{2-1} = 12x$ Differentiate using the rule for powers.

6. $g(x) = \dfrac{12x^3 - 8x}{2x}$

$$g'(x) = \frac{(2x)\frac{d}{dx}(12x^3 - 8x) - (12x^3 - 8x)\frac{d}{dx}(2x)}{(2x)^2}$$

Apply the pattern for the quotient rule.

$g'(x) = \dfrac{(2x) \cdot (36x^2 - 8) - (12x^3 - 8x) \cdot (2)}{(2x)^2}$ Differentiate.

$g'(x) = \dfrac{(72x^3 - 16x) - (24x^3 - 16x)}{4x^2}$ Simplify the numerator and denominator.

$g'(x) = \dfrac{72x^3 - 16x - 24x^3 + 16x}{4x^2}$ Clear the parentheses.

$g'(x) = \dfrac{48x^3}{4x^2} = 12x$ Combine like terms and reduce.

Chapter 6

1. The accompanying figure shows what should be typed on the home screen. Remember that "nDeriv(" is found in the catalogue or under the MATH menu, choice number 8.

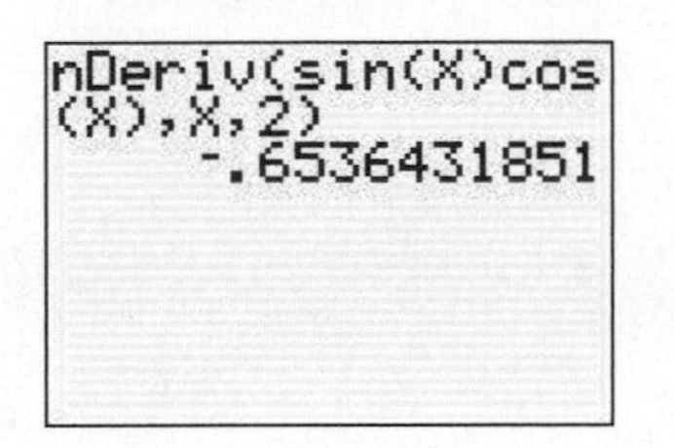

2. Figure 6-8 shows what must be typed on the "$y =$" screen and what the graph should look like.

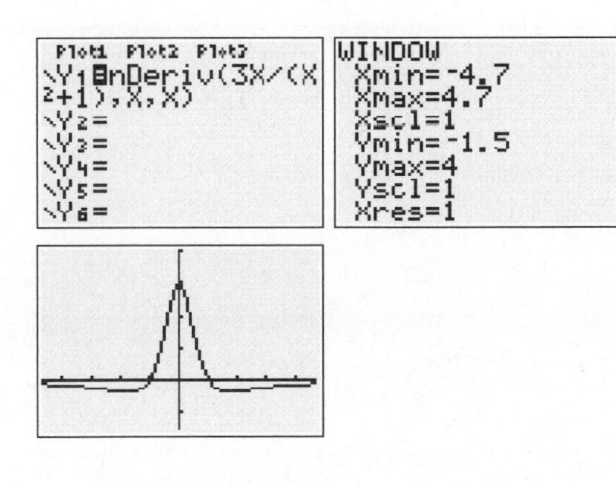

3. At $x = \frac{\pi}{2}$, $y = \cos\left(\frac{\pi}{2}\right) = 0$. The ordered pair is $\left(\frac{\pi}{2}, 0\right)$.

$$\frac{d}{dx}[\cos(x)] = -\sin(x)$$

At $x = \frac{\pi}{2}$, $-\sin\left(\frac{\pi}{2}\right) = -1$, so the slope is -1.

The point-slope form of the equation is $y - 0 = -1\left(x - \frac{\pi}{2}\right)$. The slope-intercept form is $y = -x + \frac{\pi}{2}$.

4. Simplify the expression first!

$$h(x) = \cos(x) \cdot \csc(x)$$

$$h(x) = \cos(x) \cdot \frac{1}{\sin(x)} = \cot(x)$$

$$h'(x) = -\csc^2(x)$$

5. Let $f(x) = \sec(x) = \frac{1}{\cos(x)}$.

$$f'(x) = \frac{\cos(x) \cdot \frac{d}{dx}(1) - 1 \cdot \frac{d}{dx}[\cos(x)]}{\cos^2(x)}$$

$$f'(x) = \frac{\cos(x) \cdot 0 - 1 \cdot [-\sin(x)]}{\cos^2(x)}$$

$$f'(x) = \frac{\sin(x)}{\cos^2(x)}$$

$$f'(x) = \frac{1}{\cos(x)} \cdot \frac{\sin(x)}{\cos(x)}$$

$$f'(x) = \sec(x) \cdot \tan(x)$$

6. If $y = 2 + \cos(x)$, then $\frac{dy}{dx} = -\sin(x)$. On $[0, 2\pi]$, the range of $\sin(x)$ is $-1 \leq \sin(x) \leq 1$. The slope of $y = 2 + \cos(x)$ is its derivative. Thus $\frac{dy}{dx} = -\sin(x)$ will be greatest when $\sin(x) = -1$, which occurs when $x = \frac{3\pi}{2}$ on the given domain.

Chapter 7

1. Think of $y=\sqrt[3]{x^3+6x}$ as $y=(x^3+6x)^{\left(\frac{1}{3}\right)}$.

$$\frac{dy}{dx}=\left(\frac{1}{3}\right)(x^3+6x)^{\left(\frac{1}{3}-1\right)}\cdot\frac{d}{dx}(x^3+6x)$$

Use the power rule and prepare to differentiate the base.

$$\frac{dy}{dx}=\left(\frac{1}{3}\right)(x^3+6x)^{\left(\frac{-2}{3}\right)}\cdot(3x^2+6)$$

The derivative of the base is the chain rule factor.

$$\frac{dy}{dx}=(x^3+6x)^{\left(\frac{-2}{3}\right)}\cdot(x^2+2)$$

Distribute the constant to the binomial.

2. If $p=\tan(w)$, then the derivative of p with respect to w is $\frac{dp}{dw}=\sec^2(w)$.

If $w=\cos(t)$, the derivative of w with respect to t is $\frac{dw}{dt}=-\sin(t)$.

$$\frac{dp}{dt}=\frac{dp}{dw}\cdot\frac{dw}{dt}$$

$$\frac{dp}{dt}=\sec^2(w)\cdot[-\sin(t)]$$

Substitute on the right-hand side of the equation.

$$\frac{dp}{dt}=-\sin(t)\cdot\sec^2[\cos(t)]$$

Substitute $\cos(t)$ for w.

3. If $y=\frac{f(x)}{g(x)}$, then using the quotient rule, $y'(x)=\frac{g(x)\cdot f'(x)-f(x)\cdot g'(x)}{[g(x)]^2}$.

$$y'(2)=\frac{g(2)\cdot f'(2)-f(2)\cdot g'(2)}{[g(2)]^2}$$

Replace x with 2.

$$y'(2)=\frac{-4\cdot\left(\frac{1}{3}\right)-5\cdot(-1)}{[-4]^2}=\frac{11}{48}$$

Substitute values from the table and simplify.

4. If $y=f(g(x))$, then using the chain rule, $y'(x)=f'(g(x))\cdot g'(x)$.

$y'(1)=f'(g(1))\cdot g'(1)$ Substitute 1 for x.

$$y'(1)=f'(2)\cdot g'(1)$$

From the table, replace $g(1)$ with 2.

$$y'(1)=\frac{1}{3}\cdot\frac{3}{4}=\frac{1}{4}$$

Substitute values from the table and simplify.

5. If $y^3=x^2+\sin(y)$, then use implicit differentiation.

$$3y^2\frac{dy}{dx}=2x+\cos(y)\cdot\frac{dy}{dx}$$

Every time a y term is differentiated, you get a $\frac{dy}{dx}$ factor from the chain rule.

$$3y^2\frac{dy}{dx}-\cos(y)\cdot\frac{dy}{dx}=2x$$

Isolate $\frac{dy}{dx}$ on one side of the equation.

$[3y^2 - \cos(y)]\frac{dy}{dx} = 2x$

Factor $\frac{dy}{dx}$ out as a greatest common factor.

$\frac{dy}{dx} = \frac{2x}{3y^2 - \cos(y)}$

Divide both sides by $3y^2 - \cos(y)$.

6. You need one point and the slope to write the equation of a line.

The point is $(1,2)$.

Use implicit differentiation on $x^2y + y^3 + x^3 = 13$ to get $\frac{dy}{dx}$ and find the slope.

$x^2\frac{dy}{dx} + y \cdot 2x + 3y^2\frac{dy}{dx} + 3x^2 = 0$

Product rule was used to differentiate the x^2y term.

$1^2\frac{dy}{dx} + 2 \cdot 2(1) + 3(2)^2\frac{dy}{dx} + 3(1)^2 = 0$

Substitute $(1,2)$ right away.

$\frac{dy}{dx} + 4 + 12\frac{dy}{dx} + 3 = 0$

Simplify and solve for $\frac{dy}{dx}$.

$13\frac{dy}{dx} = -7$

$\frac{dy}{dx} = \frac{-7}{13}$

$y - 2 = \frac{-7}{13}(x-1)$

Chapter 8

1. Let $y = 3^u$ and $u = x^2 - x$.

$\frac{dy}{du} = 3^u \cdot \ln(3)$ and $\frac{du}{dx} = 2x - 1$

$\frac{dy}{dx} = \frac{dy}{du} \cdot \frac{du}{dx}$

$\frac{dy}{dx} = 3^u \ln(3) \cdot (2x-1)$

Substitute expressions in for $\frac{dy}{du}$ and $\frac{du}{dx}$.

$\frac{dy}{dx} = 3^{(x^2-x)}\ln(3) \cdot (2x-1)$ Substitute for u.

2. All that you need for an equation of a line is a slope and a point. You have the point $(2,0)$, and the slope comes from the derivative. Because $y = \ln(x^3 - 7)$, use the rule for derivatives of natural logarithms, $\frac{dy}{dx} = \frac{1}{x^3-7} \cdot \frac{d}{dx}(x^3 - 7) = \frac{1}{x^3-7} \cdot 3x^2$. Now plug in 2 for x, and $\frac{dy}{dx} = \frac{1}{2^3-7} \cdot 3(2)^2 = 12$. In point-slope form, the equation of the line is $y - 0 = 12(x-2)$, or $y = 12(x-2)$.

3. Because there is a variable in the base and the exponent, logarithmic differentiation must be used.

$f(x) = x^{(x+2)}$ or $y = x^{(x+2)}$

Take the natural logarithm of both sides of the equation. $\ln(y) = \ln[x^{(x+2)}]$

Bring the exponent down in front. $\ln(y) = (x+2)\ln(x)$

Differentiate both sides of the equation.

$$\frac{1}{y}\cdot\frac{dy}{dx}=(x+2)\cdot\frac{1}{x}+\ln(x)\cdot 1$$

Move y over to the right-hand side, and replace it with $x^{(x+2)}$. $\frac{dy}{dx}=x^{(x+2)}\left[(x+2)\cdot\frac{1}{x}+\ln(x)\right]$

Now evaluate when $x=1$.

$$\frac{dy}{dx}=1^{(1+2)}\left[(1+2)\cdot\frac{1}{1}+\ln(1)\right]=3.$$

4. This is a chain rule problem, so use $\frac{dy}{dx}=\frac{dy}{du}\cdot\frac{du}{dx}$.

Rewrite $y=\log_5(u)$ as $y=\frac{\ln(u)}{\ln(5)}$. Then $\frac{dy}{du}=\frac{1}{\ln(5)}\cdot\frac{1}{u}$.

If $u=1+\sqrt{x}$, then $\frac{du}{dx}=\frac{1}{2}x^{(-1/2)}$.

Multiply the results to get $\frac{dy}{dx}=\frac{1}{\ln(5)}\cdot\frac{1}{u}\cdot\frac{1}{2}x^{(-1/2)}$.

Now substitute for u and clean up the answer a bit. $\frac{dy}{dx}=\frac{1}{2\ln(5)}\cdot\frac{1}{(1+\sqrt{x})}\cdot\frac{1}{\sqrt{x}}$

5. As complicated as this may sound, a derivative is a rate of growth, and the function is just a form of $y=a^x$ with t taking the place of x. Thus the answer is found by evaluating the derivative when $t=3$.

$A=10000(1.01)^{4t}$, so

$\frac{dA}{dt}=10000(1.01)^{4t}\ln(1.01)\cdot 4$. Remember that the 4 comes from the derivative of the exponent $4t$. Evaluate the derivative expression on your calculator, and you should get $\frac{dA}{dt}\approx 448.49$ dollars per year.

Chapter 9

1. Let $y=\cos^{-1}(u)$ and $u=3x$.

$$\frac{dy}{du}=\frac{-1}{\sqrt{1-u^2}} \text{ and } \frac{du}{dx}=3.$$

$$\frac{dy}{dx}=\frac{dy}{du}\cdot\frac{du}{dx} \text{ so } \frac{dy}{dx}=\frac{-1}{\sqrt{1-u^2}}\cdot 3.$$

Substituting for u yields

$$\frac{dy}{dx}=\frac{-3}{\sqrt{1-(3x)^2}}=\frac{-3}{\sqrt{1-9x^2}}.$$

2. Use product rule to find the derivative, and then evaluate at $x=1$.

$$f'(x)=x^4\cdot\frac{1}{1+x^2}+4x^3\cdot\tan^{-1}(x)$$

$$f'(1)=1^4\cdot\frac{1}{1+1^2}+4(1)^3\cdot\tan^{-1}(1)$$

$$f'(1)=\frac{1}{2}+4\cdot\frac{\pi}{4}=\frac{1}{2}+\pi$$

3. Find the point and the slope.

At $x=2$, $y=\sec^{-1}(2)$. The angle whose secant is 2 is the same angle whose cosine is $\frac{1}{2}$.

The ordered pair is $\left(2,\frac{\pi}{3}\right)$.

If $y=\sec^{-1}(x)$, then $\frac{dy}{dx}=\frac{1}{|x|\cdot\sqrt{x^2-1}}$.

At $x=2$, $\frac{dy}{dx}=\frac{1}{|2|\cdot\sqrt{2^2-1}}=\frac{1}{2\sqrt{3}}$.

$y-\frac{\pi}{3}=\frac{1}{2\sqrt{3}}(x-2)$

4. Let $y=\cot^{-1}(u)$ and $u=\frac{1}{x}=x^{-1}$.

$\frac{dy}{du}=\frac{-1}{1+u^2}$ and $\frac{du}{dx}=-1x^{-2}=\frac{-1}{x^2}$.

$\frac{dy}{dx}=\frac{dy}{du}\cdot\frac{du}{dx}$ so $\frac{dy}{dx}=\frac{-1}{1+u^2}\cdot\frac{-1}{x^2}$.

Substituting for u produces $\frac{dy}{dx}=\frac{-1}{1+\left(x^{-1}\right)^2}\cdot\frac{-1}{x^2}$.

Multiply the fractions and $\frac{dy}{dx}=\frac{1}{x^2+1}$. Surprisingly, this is the derivative of $y=\tan^{-1}(x)$, but maybe it's not so surprising. Recall the trigonometric identity that $\cot^{-1}\left(\frac{1}{x}\right)=\tan^{-1}(x)$. Finding the solution to the problem would have been much quicker if we had used this identity first!

5. Let $g(u)=\csc^{-1}(u)$ and $u(x)=e^{5x}$, so $g(u(x))=\csc^{-1}(e^{5x})$.

$g'(u)=\frac{-1}{|u|\cdot\sqrt{u^2-1}}$ and $u'(x)=5e^{5x}$.

$\frac{d}{dx}[g(u(x))]=g'(u(x))\cdot u'(x)$

$\frac{dg}{dx}=\frac{-1}{|u|\cdot\sqrt{u^2-1}}\cdot 5e^{5x}$

Substituting for u yields

$\frac{dg}{dx}=\frac{-1}{|e^{5x}|\cdot\sqrt{(e^{5x})^2-1}}\cdot 5e^{5x}$.

$\frac{dg}{dx}=\frac{-5}{\sqrt{e^{10x}-1}}$

Chapter 10

1. $k'(x)=12x^3+2x-3$ and $k''(x)=36x^2+2$
Find the first and second derivatives.

$\frac{k'(0)}{k''(0)}=\frac{-3}{2}$ Examine the ratios at $x=0$.

2. $h'(x)=-2e^{(-2x)}$

Use the rule $\frac{d}{dx}e^{u(x)}=e^{u(x)}\frac{du}{dx}$.

$h''(x)=-2\cdot-2e^{(-2x)}=4e^{(-2x)}$

$h''(3)=4e^{(-6)}>0$ because all real powers of e are positive. The second derivative of h is positive, so the first derivative of h is increasing.

3. $v(t)=s'(t)=-12t^2+24t+72$

$a(t)=v'(t)=s''(t)=-24t+24$

At $t=2$ seconds,
$v(2)=-12(2)^2+24(2)+72=72$ feet per second.

$a(2)=-24(2)+24=-24$ feet per second per second

The engine is still firing. As soon as the engine stops providing acceleration, the only acceleration acting on the rocket will be gravity, which is –32 feet per second per second. Because the acceleration at 2 seconds is greater than –32 feet per second per second, the rocket must still be providing some power.

4. $m'(t)$ is negative, because for any time interval where the change in time is positive, the change in mass will be negative, so their ratio will be negative.

   $m''(t)$ is positive. Because the polar ice caps are melting more and more each year, the rate of melting, which is $m'(t)$, is increasing and has positive slope; therefore $m''(t)$ is positive.

5. Use implicit differentiation on $\frac{dy}{dx}$. Remember the quotient rule.

$$\frac{dy}{dx} = \frac{x^5}{y^2+1}$$

$$\frac{d}{dx}\left(\frac{dy}{dx}\right) = \frac{(y^2+1)\cdot 5x^4 - x^5 \cdot 2y\frac{dy}{dx}}{(y^2+1)^2}$$

Use the quotient rule and get $\frac{dy}{dx}$ from differentiating the y term.

$$\frac{d^2y}{dx^2} = \frac{5x^4(y^2+1) - 2x^5y\left(\frac{x^5}{y^2+1}\right)}{(y^2+1)^2}$$

Substitute for $\frac{dy}{dx}$ from the original given information.

$$\frac{d^2y}{dx^2} = \frac{5x^4(y^2+1)^2 - 2x^{10}y}{(y^2+1)^3}$$

Multiply the numerator and denominator of the fraction by y^2+1.

Chapter 11

1. The thinner graph is $y = x^{\left(\frac{2}{3}\right)}$, and the thicker graph is its derivative. To the left of the y-axis, $y = x^{\left(\frac{2}{3}\right)}$ is always decreasing, so its derivative is always negative. Also, the slopes of the tangents are getting more negative as the graph gets closer to the y-axis, so the graph of the derivative must get more negative. To the right of the y-axis, $y = x^{\left(\frac{2}{3}\right)}$ is always increasing, but it is getting less steep as it moves away from the y-axis. As a result, for $x > 0$ its derivative is always positive but is getting smaller as x gets bigger. At the y-axis, $y = x^{\left(\frac{2}{3}\right)}$ has a cusp. Therefore, its derivative does not exist, and the graph of the derivative should have a vertical asymptote at $x = 0$.

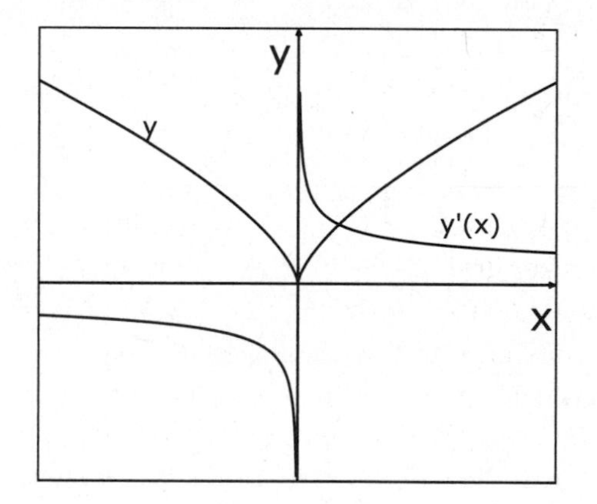

2. $g(x)$ is decreasing and concave up on the interval. If $g'(x)$ is negative on an interval, then $g(x)$ is decreasing on the interval. Since $g'(x)$ is increasing on the interval, its derivative $g''(x)$ is positive. If $g''(x) > 0$, then $g(x)$ is concave up.

3. The thicker graph is $h''(x)$. Remember that $h''(x)$ is just the derivative of $h'(x)$. The first section of $h'(x)$ is decreasing with the slopes of the tangents getting more negative, so $h''(x)$ starts negative and gets more negative. The second section of $h'(x)$ is a horizontal segment with a constant slope of zero. The third section of $h'(x)$ starts with a large positive slope and stays positive over the interval, but the slope is decreasing, so $h''(x)$ is all positive but decreasing.

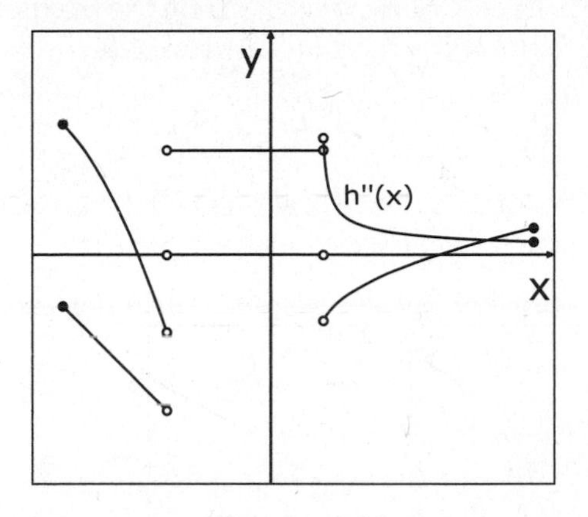

4. The thicker graph is $p(x)$. As long as your sketch has the same shape, it does not matter where $p(x)$ is oriented vertically on the coordinate plane. $p'(x)$ is defined on the entire given interval, so $p(x)$ should be differentiable at all points. The first segment of $p'(x)$ is positive and decreasing, so $p(x)$ should be increasing but concave down. The second section of $p'(x)$ is always zero, so $p(x)$ should be a horizontal segment that connects smoothly with the first section of $p(x)$. The third section of $p'(x)$ is negative and decreasing, so $p(x)$ should be decreasing and concave down.

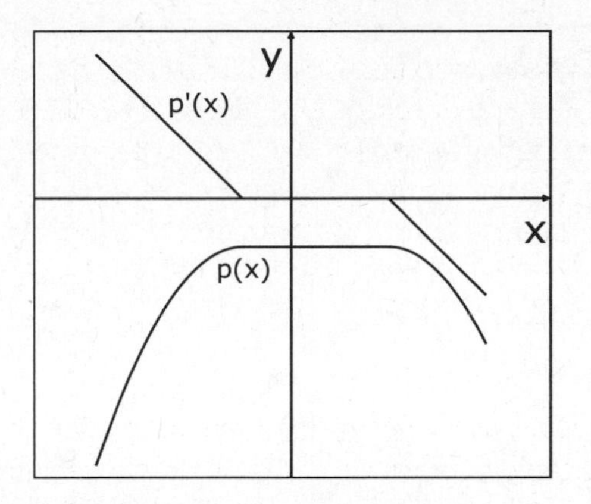

5. Put the points on the grid first. Then, on the interval $[-2,1]$, $r'(x)>0$ and $r''(x)>0$, so $r(x)$ must be increasing and concave up from (-2,0) to (1,3). On the interval $[1,3]$, $r'(x)<0$ and $r''(x)<0$, so $r(x)$ must be decreasing and concave down from (1,3) to (3,1). Because $r(x)$ must be continuous but $r'(1)$ is undefined, there must be a corner at (1,3).

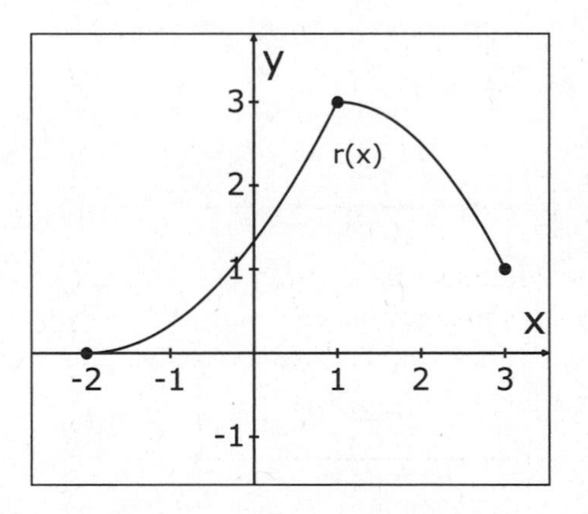

Chapter 12

1. $k'(x) = 2(x-2) = 0 \Rightarrow x = 2$

$k''(x) = 2 > 0$, so $x = 2$ is local minimum.

$k(0) = 7$, $k(2) = 3$, $k(5) = 12$

At $x = 0$, the local maximum value is 7. At $x = 2$, the local and absolute minimum value is 3. At $x = 5$, the local and absolute maximum is 12.

2. $h'(x) = \dfrac{2x}{x^2+1}$ and $h''(x) = \dfrac{2(x^2+1) - 2x(2x)}{(x^2+1)^2}$

$\dfrac{2(x^2+1) - 2x(2x)}{(x^2+1)^2} = 0$ only when the numerator equals zero. The denominator will never equal zero.

$2x^2 + 2 - 4x^2 = 0$

$2 - 2x^2 = 0 \Rightarrow x = \pm 1$

$h''(x)$ changes sign on each side of the determined x values, so $h(x)$ has inflection points at both places.

3. At $x = -5$, $w'(x)$ changes from positive to negative, so $w(x)$ goes from increasing to decreasing and has a local maximum. At $x = 1$, $w'(x)$ changes from negative to positive, so $w(x)$ goes from decreasing to increasing and has a local minimum. At $x = -2$, $w'(x)$ goes from decreasing to increasing, so $w''(x)$ goes from negative to positive and produces an inflection point at $x = -2$ on $w(x)$.

4. Let x be the side of each square cut out from the corners. The dimensions of the completed box will be x by $6 - 2x$ by $6 - 2x$.

$V(x) = x(6-2x)^2 = 4x^3 - 24x^2 + 36x$

$V'(x) = 12x^2 - 48x + 36 = 0$

$12(x-1)(x-3) = 0 \Rightarrow x = 1$ or $x = 3$

$V''(x) = 24x - 48$

$V''(3) = 72 - 48 > 0$ makes $x = 3$ a minimum.

$V''(1) = 24 - 48 < 0$, which means that $x = 1$ provides a maximum. The dimensions are 1 in. by 4 in. by 4 in.

5. $f(x) = \dfrac{2^x}{x}$

$f(1) = \dfrac{2^1}{1} = 2$ and $f(4) = \dfrac{2^4}{4} = 4$, so the points are $(1,2)$ and $(4,4)$.

$\dfrac{2^c(c\ln(2) - 1)}{c^2} = \dfrac{4-2}{4-1}$

Set up according to the Mean Value Theorem formula.

$c \approx 2.506$

Solve by graphing the left-hand side of the equation as y1 and the right-hand side as y2 and intersecting the graphs.

6. $v(t) = x'(t) = t^2 - 9$

$a(t) = v'(t) = 2t$

$v(t) = t^2 - 9 = 0 \Rightarrow t = 3$

$x(3) = \frac{1}{3}(3)^3 - 9(3) + 1 = -17$ and

$a(3) = 2(3) = 6$

7.

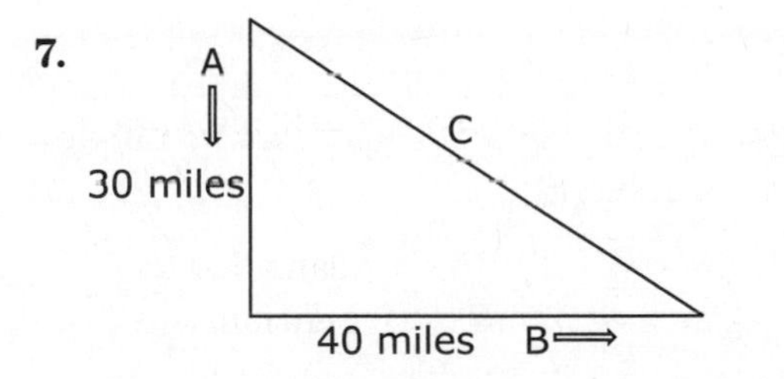

$A = 30$ and $\frac{dA}{dt} = -40$ because the distance to the station is decreasing.

$B = 40$ and $\frac{dB}{dt} = 60$ because the distance from the station is increasing.

By the Pythagorean Theorem $A^2 + B^2 = C^2$, so $C = 50$ miles. You are seeking $\frac{dC}{dt}$.

$2A\frac{dA}{dt} + 2B\frac{dB}{dt} = 2C\frac{dC}{dt}$

Differentiate the Pythagorean Theorem with respect to time.

$2(30)(-40) + 2(40)(60) = 2(50)\frac{dC}{dt}$

Substitute known values.

$100\frac{dC}{dt} = 2400$

$\frac{dC}{dt} = 24$ miles per hour

Chapter 13

1. Find the width of each rectangle, and multiply by the height of each of five rectangles whose height is based on the right-hand side of each interval.

$$\Delta x = \frac{2\pi - \pi}{5} = \frac{\pi}{5}$$

$$\text{RRAM}_5 = \frac{\pi}{5}\left[h\left(\frac{6\pi}{5}\right) + h\left(\frac{7\pi}{5}\right) + h\left(\frac{8\pi}{5}\right) + h\left(\frac{9\pi}{5}\right) + h(2\pi)\right] \approx 17.377$$

2. The four subintervals are [0,1], [1,2], [2,3] and [3,4], so the heights of the rectangles are based on $g(x)$ evaluated at the midpoint of each interval.

$$\Delta x = \frac{4-0}{4} = 1$$

$$\text{MRAM}_4 = 1 \cdot \left[g\left(\frac{1}{2}\right) + g\left(\frac{3}{2}\right) + g\left(\frac{5}{2}\right) + g\left(\frac{7}{2}\right)\right] \approx 6.801$$

3. Find the width of each rectangle, and multiply by the height of each of six rectangles whose height is based on the left-hand side of each interval.

$$\Delta x = \frac{5-2}{6} = \frac{1}{2}$$

$$\text{LRAM}_6 = \frac{1}{2}[k(2) + k(2.5) + k(3) + k(3.5) + k(4) + k(4.5)] \approx 2.814$$

4. The left Riemann sum will use the speed at the left end of each interval, and therefore it will not use the last speed entry. The change in time for each interval is the same, 3 seconds. Accumulated distance is the sum of "rate times time" for each interval.
Distance $\approx 28 \cdot 3 + 50 \cdot 3 + 70 \cdot 3 + 88 \cdot 3 = 708$ feet

5. $\int_3^8 \frac{x}{x^2 - 1}dx = \lim_{n\to\infty} \sum_{i=1}^{n} \frac{x_i}{(x_i)^2 - 1}\Delta x$ on the interval [3,8]

6. Find the area under $f(x) = \sqrt{x+5}$ on the interval [3,9], using four trapezoids and four divisions of Simpson's rule. Compare your results to the area accurate to five decimal places, which is 19.83719.

In each case, $\Delta x = \frac{9-3}{4} = \frac{3}{2}$.

For the trapezoidal rule, $\frac{\Delta x}{2} = \frac{3}{4}$.

$$\text{TRAP}_4 = \frac{3}{4}\left[f(3) + 2f\left(\frac{9}{2}\right) + 2f(6) + 2f\left(\frac{15}{2}\right) + f(9)\right] = 19.82911$$

For Simpson's rule, $\frac{\Delta x}{3} = \frac{1}{2}$.

$$\text{SIMP}_4 = \frac{1}{2}\left[f(3) + 4f\left(\frac{9}{2}\right) + 2f(6) + 4f\left(\frac{15}{2}\right) + f(9)\right] = 19.83715$$

With just four subdivisions, Simpson's rule is accurate to the fourth decimal place, whereas the trapezoidal rule is accurate only to the first decimal place.

7.

```
fnInt(2+√(X+3),X
,-1,3)
         15.91234089
```

Chapter 14

1. The graph of $y = x^2 - 6x$ lies below the x-axis on the interval [1,5], so the definite integral will be negative. To compute area, simply find the opposite of the integral. On your calculator, type $-\text{fnInt}(x^2 - 5x,\ x,\ 1,\ 5)$. The value is $30\frac{2}{3}$.

2a. $\int_1^4 (3 + \sqrt{x})\,dx = \int_1^4 3\,dx + \int_1^4 \sqrt{x}\,dx$

$$\int_1^4 3\,dx + \int_1^4 \sqrt{x}\,dx = 3(4-1) + \frac{14}{3} = \frac{41}{3}$$

2b. $\int_4^1 6\sqrt{x}\,dx = -6\int_1^4 \sqrt{x}\,dx$

$$-6\int_1^4 \sqrt{x}\,dx = -6\cdot\frac{14}{3} = -28$$

2c. $\int_4^1 (3\sqrt{x}-5)\,dx = -\int_1^4 (3\sqrt{x}-5)\,dx$

$$-\int_1^4 (3\sqrt{x}-5)\,dx = -3\int_1^4 \sqrt{x}\,dx + \int_1^4 5\,dx$$

$$-3\int_1^4 \sqrt{x}\,dx + \int_1^4 5\,dx = -3\cdot\frac{14}{3} + 5(4-1) = 1$$

3. Set up all integrals so that the limits are ordered from left to right. Split up sums. Factor the constants out of the integrals. The most subtle result is $\int_2^3 h(x)\,dx = -B$. The function lies below the x-axis on the interval, so the definite integral must be negative.

$$\int_2^0 [3h(x)+5]dx + \int_2^3 4h(x)\,dx = -3\int_0^2 h(x) - \int_0^2 5\,dx + 4\int_2^3 h(x)\,dx$$

$$-3\int_0^2 h(x) - \int_0^2 5\,dx + 4\int_2^3 h(x)\,dx = -3A - 5(2-0) + 4(-B)$$

$$-3\mathrm{A} - 5(2-0) + 4(-B) = -3\mathrm{A} - 4B - 10$$

4. If $\int_a^b p(x)\,dx = 3a - 7b$, then find an expression in terms of a and b for $\int_b^a [p(x)+8]dx$.

$$\int_b^a [p(x)+8]dx = -\int_a^b [p(x)+8]dx$$

$$-\int_a^b [p(x)+8]dx = -\int_a^b p(x)\,dx - \int_a^b 8\,dx$$

$$-\int_a^b p(x)\,dx - \int_a^b 8\,dx = -(3a-7b) - 8(b-a)$$

$$-(3a-7b) - 8(b-a) = -3a + 7b - 8b + 8a = 5a - b$$

5. The absolute value function makes the entire graph positive. Make sure your calculator is in radian mode. On the Home screen, type fnInt(abs($x\sin(2x)$), x, 0, 2π), as shown. Remember that the absolute value command is in the catalogue. Accurate to three decimal places, the result is 12.566.

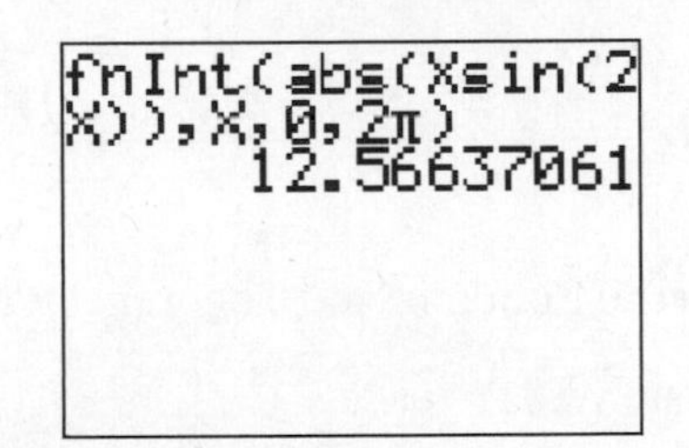

Chapter 15

1. By the first part of the Fundamental Theorem, if $f(x)=\int_2^x \sec(t)\,dx$, then

$f'(x)=\sec(x)$ and $f'\left(\frac{\pi}{3}\right)=\sec\left(\frac{\pi}{3}\right)$. Because $\sec(x)=\frac{1}{\cos(x)}$, $\sec\left(\frac{\pi}{3}\right)=\frac{1}{1/2}=2$.

2. The upper limit is a composite function of x, so you need to use the chain rule in conjunction with the first part of the Fundamental Theorem. The short method is to replace every t with x^4. $\frac{d}{dx}\int_3^{x^4}\frac{t}{\sqrt{7+t}}dt=\frac{x^4}{\sqrt{7+x^4}}d(x^4)$.

$$\frac{d}{dx}\int_3^{x^4}\frac{t}{\sqrt{7+t}}dt=\frac{x^4}{\sqrt{7+x^4}}4x^3=\frac{4x^7}{\sqrt{7+x^4}}$$

The other way to do this problem is to write $y=\int_3^u\frac{t}{\sqrt{7+t}}dt$ and $u=x^4$. Then

$\frac{dy}{dx}=\frac{dy}{du}\cdot\frac{du}{dx}$. If you prefer this and forget the process, see Example 15-6.

3. Remember that the equation of a tangent line requires a point and a slope. Find $h(1)$ to get the point and $h'(1)$ to get the slope. Because $h(x)=\int_{x^2}^{x^3}2^{(t-1)}\,dt$,

$h(1)=\int_1^1 2^{(t-1)}\,dt=0$. The point is $(1,0)$. Following the model of Example 15-7, $h'(x)=2^{(x^3-1)}\cdot 3x^2-2^{(x^2-1)}\cdot 2x$. Then $h'(1)=2^0\cdot 3-2^0\cdot 2=1$. In point-slope form, the equation of the tangent line is $y-0=1(x-1)$.

4a. Because $r(x)=\int_{-1}^{x} p(t)\,dt$, $r(4)=\int_{-1}^{4} p(t)\,dt$, which is the net area between the graph of $p(t)$ and the x-axis. Using geometry, $\int_{-1}^{0} p(t)\,dt=-\frac{1}{2}$ and $\int_{0}^{4} p(t)\,dt=\frac{1}{2}\cdot 4\cdot 2=4$. Summing the results, you get

$$r(4)=\int_{-1}^{4} p(t)\,dt=3\frac{1}{2}.$$

4b. Given that $r(x)=\int_{-1}^{x} p(t)\,dt$, by the Fundamental Theorem, $r'(x)=p(x)$. Reading right off the graph of p reveals that $r'(3)=p(3)=1$.

4c. From part b, $r'(x)=p(x)$, so $r''(x)=p'(x)$. To find where $r(x)$ is concave down, find where $r''(x)<0$, which implies $p'(x)<0$, which means that p is decreasing. From the figure, p is decreasing on the intervals $(-3,-1)$ and $(2,4)$.

5. For $\int_{1}^{2}(x^3+2x-3)\,dx$, the antiderivative is $G(x)=\frac{1}{4}x^4+x^2-3x$.

The $\frac{1}{4}$ balances the exponent of 4, which is brought down when differentiating. Using the second part of the Fundamental Theorem yields

$$G(2)-G(1)=\left[\frac{1}{4}(2)^4+2^2-3(2)\right]-\left[\frac{1}{4}(1)^4+1^2-3(1)\right]=3\frac{3}{4}.$$

6. Because $\cos(x)$ is the derivative of $\sin(x)$, the antiderivative for $\int_{0}^{\pi/2} 3\cos(x)\,dx$ is $G(x)=3\sin(x)$.

$$G\left(\frac{\pi}{2}\right)-G(0)=3\sin\left(\frac{\pi}{2}\right)-3\sin(0)=3$$

Chapter 16

1. Simplify the integrand algebraically.

$$\int_3^5 \frac{x^2-4}{x-2}\,dx = \int_3^5 \frac{(x-2)(x+2)}{x-2}\,dx$$

$$\int_3^5 \frac{(x-2)(x+2)}{x-2}\,dx = \int_3^5 (x+2)\,dx$$

$$\int_3^5 (x+2)\,dx = \frac{1}{2}x^2 + 2x\Big|_3^5$$

$$\frac{1}{2}x^2 + 2x\Big|_3^5 = \left(\frac{25}{2}+10\right) - \left(\frac{9}{2}+6\right) = 12$$

2. Take a geometric approach. $y = -\sqrt{16-x^2}$ is a semicircle under the x-axis, centered at the origin, with a radius of 4.

$$\int_{-4}^{4} -\sqrt{16-x^2}\,dx = -\frac{1}{2}\pi\cdot 4^2 = -8\pi$$

3. Distribute the x^2 term and integrate term by term.

$$\int_1^3 x^2\left(3+\frac{2}{x^3}\right)dx = \int_1^3 \left(3x^2+\frac{2}{x}\right)dx$$

$$\int_1^3 \left(3x^2+\frac{2}{x}\right)dx = x^3 + 2\ln(x)\Big|_1^3$$

$$x^3 + 2\ln(x)\Big|_1^3 = [3^3 + 2\ln(3)] - [1^3 + 2\ln(1)] = 26 + 2\ln(3)$$

4. $\int_0^{2\pi} \cos(x)\,dx = -\sin(x)\big|_0^{2\pi} = -[\sin(2\pi) - \sin(0)] = 0$

The graph of $y = \cos(x)$ on the interval $[0, 2\pi]$ has an equal amount of area above and below the x-axis, so taking a geometric approach, you would expect the value of the definite integral to be 0.

5. Because x is a multiple of the derivative of $x^2 - 1$, use u-substitution.

$$\int_1^3 x \cdot \sqrt[3]{x^2 - 1}\, dx$$

Let $u = x^2 - 1$ and $du = 2x\, dx$, so $x\, dx = \frac{1}{2} du$.

Change the limits as well. $x = 1 \Rightarrow u = 0$, and $x = 3 \Rightarrow u = 8$.

$$\int_1^3 x \cdot \sqrt[3]{x^2 - 1}\, dx = \int_0^8 u^{(1/3)} \cdot \frac{1}{2} du$$

$$\frac{1}{2}\int_0^8 u^{(1/3)}\, du = \frac{1}{2} \cdot \frac{3}{4} u^{(4/3)}\Big|_0^8$$

$$\frac{1}{2} \cdot \frac{3}{4} u^{(4/3)}\Big|_0^8 = \frac{3}{8}\left(8^{(4/3)} - 0\right) = 6$$

6. This also lends itself to u-substitution. Recall that the derivative of $\tan^{-1}(x)$ is $\frac{1}{1+x^2}$.

$$\int_0^1 \frac{\tan^{-1}(x)}{1+x^2} dx$$

Let $u = \tan^{-1}(x)$ and $du = \frac{1}{1+x^2} dx$.

Change the limits. $x = 0 \Rightarrow u = 0$, and $x = 1 \Rightarrow u = \tan^{-1}(1) = \frac{\pi}{4}$.

$$\int_0^1 \frac{\tan^{-1}(x)}{1+x^2} dx = \int_0^{\pi/4} u\, du$$

$$\int_0^{\pi/4} u\, du = \frac{1}{2} u^2\Big|_0^{\pi/4} = \frac{\pi^2}{32}$$

7. $\int_1^2 x \cdot e^{2x}\, dx$ requires integration by parts.

Using LIPET, let $u = x$ and $dv = e^{2x}\, dx$.

Then $du = dx$ and $v = \frac{1}{2} e^{2x}$. The $\frac{1}{2}$ accounts for the chain rule factor of 2 when you are differentiating.

$$uv - \int v\, du = \frac{1}{2} x e^{2x}\Big|_1^2 - \int_1^2 \frac{1}{2} e^{2x}\, dx$$

$$\frac{1}{2} x e^{2x}\Big|_1^2 - \int_1^2 \frac{1}{2} e^{2x}\, dx = \left(\frac{1}{2} x e^{2x} - \frac{1}{4} e^{2x}\right)\Big|_1^2$$

$$\left(\frac{1}{2} x e^{2x} - \frac{1}{4} e^{2x}\right)\Big|_1^2 = \frac{3}{4} e^4 - \frac{1}{4} e^2$$

Chapter 17

1. $\frac{df}{dx} = 3 + e^{1-x}$ and $f(1) = 7$.

$df = [3 + e^{1-x}]dx$ Separate the variables.

$$\int df = \int [3 + e^{1-x}]dx$$

Integrate each term individually. If necessary, for $\int e^{1-x}\, dx$, let $u = 1 - x$.

$$f(x) = 3x - e^{1-x} + C$$

$$7 = 3 - e^{(1-1)} + C \Rightarrow C = 5$$

2. Velocity is the derivative of position, so start with $\frac{ds}{dt} = -9.8t$, where $s(t)$ is the position above the ground at any time t.

$\int ds = \int -9.8t$

Separate the variables and integrate.

$s(t) = -4.9t^2 + C$

$100 = -4.9(0)^2 + C \Rightarrow C = 100$

Use the initial condition.

$s(t) = -4.9t^2 + 100$

$s(t) = 0$ on the ground, so

$0 = -4.9t^2 + 100 \Rightarrow t \approx 4.518$ seconds.

3. $\dfrac{dy}{dx} = xy$

$\int \dfrac{dy}{y} = \int x\,dx$ Separate and integrate.

$\ln|y| = \dfrac{1}{2}x^2 + C$

The absolute value is used because natural logarithms are not defined for negative values.

$|y| = e^{\left(\frac{1}{2}x^2 + C\right)} = e^C \cdot e^{\left(\frac{1}{2}x^2\right)}$

$y = Ae^{\left(\frac{1}{2}x^2\right)}$ where $A = \pm e^C$.

The value of A is determined by the initial condition.

4. Because $\dfrac{dy}{dx} = xy$, at any point where $x = 0$ or $y = 0$, the slope mark will be horizontal. At other points, multiply the x- and y-coordinates to get the slope. For example, at $(-1,2)$, the slope is -2.

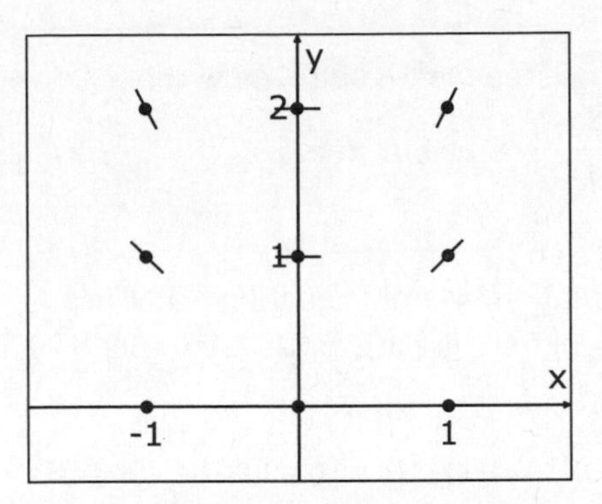

In Problems 3 and 4, combined, you found an equation of a solution curve and the slopefield. The accompanying figure shows a detailed slope-field with one of the solution curves plotted on it.

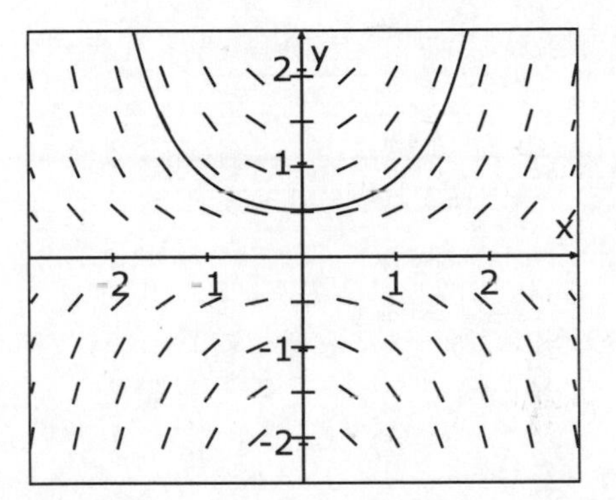

5. The decay is modeled by an exponential function; call it $A = A_0 \cdot e^{kt}$. To find k, you need to realize that in 10 years, there is 96% of the original amount left.

$0.96A_0 = A_0 \cdot e^{10k} \Rightarrow e^{10k} = 0.96$

$\ln(e^{10k}) = \ln(0.96)$

$10k = \ln(0.96) \Rightarrow k \approx -0.00408$

If $A = A_0 \cdot e^{-.00408t}$, then 10% of the initial amount will be left when $0.10A_0 = A_0 \cdot e^{-0.00408t}$.

Repeat the steps to find k to show that $t = \frac{\ln(0.10)}{-0.00408} \approx 564$ years.

6. The original differential equation is not in a proper form for comparison, so factor out $0.0003P$.

$$\frac{dP}{dt} = 0.72P - 0.0003P^2 = 0.0003P(2400 - P)$$

By comparison, $M = 2400$ and $\frac{k}{M} = 0.0003$, so $k = (0.0003)(2400) = 0.72$.

Up to this point, $P = \frac{2400}{1 + Ae^{(0.72t)}}$. Using $P(0) = 16$, $16 = \frac{2400}{1 + Ae^0} \Rightarrow A = 149$.

$$P = \frac{2400}{1 + 149e^{(0.72t)}}$$

Chapter 18

1. $\int_3^6 c(t)\,dt = 51.9$ From 3 P.M. to 6 P.M. a total of 51.9 hundred, or 5190, cars passed through the toll plaza.

2. Integrate $\frac{dH}{dt} = -10e^{\left(\frac{-t}{13}\right)}$ from 0 to 5 to determine the total change in temperature.

$$\int_0^5 -10e^{\left(\frac{-t}{13}\right)}\,dt = -10 \cdot -13e^{\left(\frac{-t}{13}\right)} = 130e^{\left(\frac{-t}{13}\right)}\Bigg|_0^5$$

Use $\int e^u\,du = e^u$ and let $u = \frac{-t}{13}$.

$$130e^{\left(\frac{-t}{13}\right)}\Bigg|_0^5 = 130e^{\left(\frac{-5}{13}\right)} - 130e^{\left(\frac{0}{13}\right)} \approx -41.5°\text{F}$$

$185°\text{F} - 41.5°\text{F} = 143.5°\text{F}$ A total decrease in temperature to the starting temperature.

3. $f_{\text{avg}} = \frac{1}{3-1}\int_0^3 e^x - 4x\,dx = \frac{1}{3}\left[e^x - 2x^2\right]_0^3$

$$\frac{1}{3}\left[e^x - 2x^2\right]_0^3 = \frac{1}{3}\left[(e^3 - 2\cdot 3^2) - (e^0 - 0)\right] \approx 0.362$$

4. Although it is close, there is more area above the x-axis on the interval $[1,3]$ than below the x-axis outside that interval. Because the net area is positive, the average value is positive.

5. The problem is seeking displacement from its starting position, not total distance, so integrate from 0 to 2 seconds and interpret the result.

$$\text{Displacement} = \int_0^{2.3} \frac{1}{2} t \cdot \sin(t^2 + 1)\,dt$$

Let $u = t^2 + 1$ and $du = 2t\,dt$.

$$\text{Displacement} = \int_1^{6.29} \frac{1}{4}\sin(u)\,du = -\frac{1}{4}\cos(u)\Bigg|_0^{6.29}$$

$$-\frac{1}{4}[\cos(6.29) - \cos(1)] \approx -0.115$$

Position at $t = 2.3$ is $3 + (-0.115) = 2.885$, or $(0, 2.885)$

6. To get total distance by hand, you need to determine where velocity is positive and where it is negative. This can be done using a sign chart or the graph of velocity. Begin by letting $v(t)=0$.

$$3t^2-8t+4=0$$

$$(3t-2)(t-2)=0$$

$$t=\frac{2}{3} \text{ or } t=2$$

Because the velocity function is a quadratic that opens upward, it is negative between its two roots and positive elsewhere.

Total distance in feet is $\int_0^{\frac{2}{3}}(3t^2-8t+4)\,dt-\int_{\frac{2}{3}}^{2}(3t^2-8t+4)\,dt+\int_2^4(3t^2-8t+4)\,dt$.

The interval with negative velocity was made positive by subtracting it. An alternative method would have been to add the integral with the upper and lower limits switched.

Chapter 19

1. Graphing each function and using the intersection tool from the CALC menu reveal the points intersection to be $x=0$ and $x=2$.

$$\int_0^2\left(\frac{5x}{1+x^2}-\frac{1}{2}x^2\right)dx=\frac{5}{2}\ln(1+x^2)-\frac{1}{2}\cdot\frac{1}{3}x^3\Big|_0^2$$

$$\frac{5}{2}\ln(1+x^2)-\frac{1}{2}\cdot\frac{1}{3}x^3\Big|_0^2=\frac{5}{2}\ln(5)-\frac{4}{3}\approx 2.69$$

2. Either observing the graph or applying algebraic methods reveals the points of intersection to be $x=0$, $x=1$, and $x=4$.

The definition of the distance between functions changes part of the way through the interval. On $[0,1]$ the height of each rectangle is $p(x)-h(x)$, and on $[1,4]$ the height of each rectangle is $r(x)-h(x)$. Two integrals are required.

$$A=\int_0^1 x^{(2/3)}-(-\sqrt{x})\,dx+\int_1^4(2-x)-(-\sqrt{x})\,dx$$

$$A=\left(\frac{3}{5}x^{(5/3)}+\frac{2}{3}x^{(3/2)}\right)\Bigg|_0^1+\left(2x-\frac{1}{2}x^2+\frac{2}{3}x^{(3/2)}\right)\Bigg|_1^4$$

$$A=\left(\frac{3}{5}+\frac{2}{3}\right)+\left(2\cdot 4+\frac{1}{2}\cdot 16+\frac{2}{3}\cdot 4^{(3/2)}\right)-\left(2\cdot 1+\frac{1}{2}\cdot 1+\frac{2}{3}\cdot 1^{(3/2)}\right)=\frac{133}{30}$$

3. Use the area formula for an equilateral triangle, $A=\frac{s^2}{4}\sqrt{3}$, where s is the length of one side. In this case, the side length is determined by the function $g(x)=3-x$. Remember that the volume is calculated by $\int_a^b A(x)\,dx$.

$$V=\int_0^2\frac{(4-x)^2}{4}\sqrt{3}\,dx=\frac{\sqrt{3}}{4}\int_0^2(16-8x+x^2)\,dx$$

$$\frac{\sqrt{3}}{4}\int_0^2(16-8x+x^2)\,dx=\frac{\sqrt{3}}{4}\cdot\left(16x-4x^2+\frac{1}{3}x^3\right)\Bigg|_0^2$$

$$\frac{\sqrt{3}}{4}\cdot\left(16x-4x^2+\frac{1}{3}x^3\right)\Bigg|_0^2=\frac{\sqrt{3}}{4}\cdot\left(32-16+\frac{8}{3}\right)=\frac{14\sqrt{3}}{3}$$

4. The region described is adjacent to the x-axis, so the disc method can be used. The radius of each disc is defined by the function.

$$V=\pi\int_1^e\left(\frac{2\ln(x)}{\sqrt{x}}\right)^2dx=\pi\int_1^e\frac{4\ln^2 x}{x}dx \quad \text{Let } u=\ln(x) \text{ and } du=\frac{1}{x}dx.$$

$$\pi \int_1^e \frac{4 \ln^2 x}{x} dx = \pi \int_{u=0}^1 4 \cdot u^2 \, du$$ If $x = 1$, then $u = 0$, and if $x = e$, then $u = 1$.

$$\pi \int_{u=0}^1 4 \cdot u^2 \, du = \pi \cdot \frac{4}{3} u^3 \Big|_0^1 = \frac{4\pi}{3}$$

5. The enclosed region is above the parabola and below the line. There is space between the region and the x-axis, so the washer method is needed. The outer radius is defined by the line and the inner radius by the parabola. Set $k(x)$ equal to $f(x)$, or graph to find that the points of intersection are at $x = -1$ and $x = 2$.

$$V = \pi \int_{-1}^2 [(x+3)^2 - (x^2+1)^2] \, dx$$

$$\pi \int_{-1}^2 [(x+3)^2 - (x^2+1)^2] \, dx = \pi \int_{-1}^2 (-x^4 - x^2 + 6x + 8) \, dx$$

$$\pi \int_{-1}^2 (-x^4 - x^2 + 6x + 8) \, dx = \pi \cdot \left(\frac{-1}{5} x^5 - \frac{1}{3} x^3 + 3x^2 + 8x \right) \Bigg|_{-1}^2$$

$$\pi \cdot \left(\frac{-1}{5} x^5 - \frac{1}{3} x^3 + 3x^2 + 8x \right) \Bigg|_{-1}^2 = \frac{117}{5}$$

6. $\frac{dy}{dx} = 2\cos(2x)$ and $\left(\frac{dy}{dx}\right)^2 = 4\cos^2(2x)$

$$L = \int_0^5 \sqrt{1 + 4\cos^2(2x)} \, dx \approx 8.521$$

Final Exam

1. False. The sign of the second derivative must change, and the original function must have a tangent line for an inflection point to exist. It is even possible for the first and second derivatives to be undefined and an inflection point still exist, as in the case of $y=\sqrt[3]{x}$ at $x=0$.

2. True. Differentiability at a point on a graph guarantees continuity.

3. False. Continuity does not guarantee differentiability. Any graph that is continuous at a point but has a corner at that point is not differentiable there.

4. False. This is the converse of the second-derivative test for local extrema, and a converse is not always true. If $h(x)$ has a cusp at $[b,h(b)]$, then there can be a local maximum, but $h'(b)$ does not exist.

5. False. For example, $\int \frac{du}{1+u^2}=\tan^{-1}(u)$. The correct equation is $\int \frac{g'(u)\,du}{g(u)}=\ln|g(u)|+C$.

6. False. The average rate of change of a function $f(x)$ on the interval $[a,b]$ is calculated by $\frac{f(b)-f(a)}{b-a}$. It is the slope of the secant between endpoints of the interval.

7. False. The total area between a function $p(x)$ and the x-axis on an interval $[a,b]$ is $\int_a^b p(x)\,dx$ only when $p(x)\geq 0$ on the entire interval.

8. True. The definite integral of velocity over a given interval calculates displacement. The definite integral of the absolute value of velocity calculates total distance.

9. True. The Intermediate Value Theorem guarantees that $k(x)$ will take on all values between –7 and 4, which includes 0, at least once in the interval.

10. True! Calculus is essentially the study of change.

11. $\lim_{x\to 5}\frac{4x^2-20x}{3x-15}=\lim_{x\to 5}\frac{4x(x-5)}{3(x-5)}=\lim_{x\to 5}\frac{4x}{3}=\frac{20}{3}$

12. Split it into two limits as shown here. In the first limit, the numerator and denominator are equal degrees but opposite in sign. In the second limit, cosine is never larger than 1, so the fraction goes to zero as the denominator gets large.

$$\lim_{x\to-\infty}\frac{7x^3+4\cos(x)}{|2x^3|}=\lim_{x\to-\infty}\frac{7x^3}{|2x^3|}+\lim_{x\to-\infty}\frac{4\cos(2x)}{|2x^3|}$$

$$\lim_{x\to-\infty}\frac{7x^3}{|2x^3|}+\lim_{x\to-\infty}\frac{4\cos(2x)}{|2x^3|}=-\frac{7}{2}+0=-\frac{7}{2}$$

13. Simply use the left branch of the piecewise function. The point does not need to exist for the limit to exist.

$$\lim_{x\to 4^-}\sin\left(\frac{\pi}{x}\right)=\sin\left(\frac{\pi}{4}\right)=\frac{\sqrt{2}}{2}$$

14. Yes, $r(x)$ is continuous at $x=-2$.

The limits from both sides of $x=-2$ must be equal and equal the value of the function at $x=-2$.

$$r(-2)=(-2)^2-(-2)-3=3$$

$$\lim_{r\to-2^-}\sqrt{7-x}=\lim_{r\to-2^+}x^2-x-3$$

$$\sqrt{7-(-2)}=(-2)^2-(-2)-3$$

$$3=3$$

15. $\frac{s(12)-s(4)}{12-4}=\frac{12\sqrt{25}-4\sqrt{9}}{12-4}=\frac{60-12}{8}=6$

16. Compare $\lim_{h\to 0}\frac{2^{(3+h)}-2^3}{h}$ to $\lim_{h\to 0}\frac{f(x+h)-f(x)}{h}$. Thus $f(x)=2^x$, and 3 has been substituted for x. When $f(x)=2^x$, then $f'(x)=2^x\ln(2)$ and $f'(3)=2^3\ln(2)=8\ln(2)$.

17. $f(x)$ must be continuous, and the derivatives from each side of 2 must exist.

$$\lim_{x\to 2^+} ax^2-bx=\lim_{x\to 2^-} ax^3+b+1 \qquad 4a-2b=8a+b+1$$

$$\lim_{x\to 2^+}\frac{d}{dx}(ax^2-bx)=\lim_{x\to 2^-}\frac{d}{dx}(ax^3+b+1)$$

$$\lim_{x\to 2^+} 2ax-b=\lim_{x\to 2^-} 3ax^2 \Rightarrow 4a-b=12a$$

Solve the system of equations $4a-2b=8a+b+1$ and $4a-b=12a$.

$a=\frac{1}{20}$ and $b=-\frac{8}{20}$.

18. $x\approx -2.246$

$y\left(\frac{\pi}{3}\right)=\frac{\sqrt{3}}{2}$ and $\frac{dy}{dx}=\cos(x)|_{x=\frac{\pi}{3}}=\cos\left(\frac{\pi}{3}\right)=\frac{1}{2}$.

The tangent line is $y-\frac{\sqrt{3}}{2}=\frac{1}{2}\left(x-\frac{\pi}{3}\right)$ or $y=\frac{1}{2}\left(x-\frac{\pi}{3}\right)+\frac{\sqrt{3}}{2}$.

19. If $f(x)=\csc(3x)$, $f'(x)=-\csc(3x)\cdot\cot(3x)\cdot\frac{d}{dx}(3x)=-3\csc(3x)\cdot\cot(3x)$.

20. $g'\left(\frac{\pi}{2}\right)=\frac{1}{4}$ Use the quotient rule, and then evaluate at $x=\frac{\pi}{2}$.

$$\frac{d}{dx}\frac{\sin(x)}{2+\cos(x)}=\frac{[2+\cos(x)]\frac{d}{dx}\sin(x)-\sin(x)\frac{d}{dx}[2+\cos(x)]}{[2+\cos(x)]^2}$$
$$=\frac{[2+\cos(x)]\cos(x)-\sin(x)\cdot-\sin(x)}{[2+\cos(x)]^2}$$

At $x=\frac{\pi}{2}$, $\cos\left(\frac{\pi}{2}\right)=0$, so $\frac{[2+\cos(x)]\cos(x)-\sin(x)\cdot-\sin(x)}{[2+\cos(x)]^2}=\frac{[2+0]0-(1\cdot-1)}{[2+0]^2}=\frac{1}{4}$.

21. $\frac{dy}{du}=\cos(u^2)\cdot 2u$ and $\frac{du}{dx}=3\sec^2(3x)$.

$$\frac{dy}{dx}=\frac{dy}{du}\cdot\frac{du}{dx}=2u\cos(u^2)\cdot 3\sec^2(3x)$$

$$\frac{dy}{dx}=2\tan(3x)\cos[\tan^2(3x)]\cdot 3\sec^2(3x)$$

$$\frac{dy}{dx}=6\sec^2(3x)\tan(3x)\cos[\tan^2(3x)]$$

22. The slope of the normal is 6.

Let $x=0$ to find the y-intercept. $0^2+\sin(0y)+y^3=8 \Rightarrow y=2$

$$\frac{d}{dx}(x^2)+\frac{d}{dx}[\sin(xy)]+\frac{d}{dx}(y^3)=\frac{d}{dx}(8)$$

$2x+\cos(xy)[x\frac{dy}{dx}+1\cdot y]+3y^2\frac{dy}{dx}=0$ The product rule was used on the angle.

Substitute $(0,2)$ to get $2(0)+\cos(0)[0\frac{dy}{dx}+1\cdot 2]+3(2)^2\frac{dy}{dx}=0$.

$$1\cdot[0+2]+12\frac{dy}{dx}=0 \Rightarrow \frac{dy}{dx}=-\frac{1}{6}$$

Find the opposite reciprocal of the tangent slope because the tangent and normal are perpendicular.

23. $p'(0)=5$

Use the product and chain rules to find the derivative of $p(t)=(4t+1)^{(1/2)}\cdot e^{3t}$.

$$p'(t)=\frac{1}{2}(4t+1)^{(-1/2)}\cdot 4\cdot e^{3t}+(4t+1)^{(1/2)}\cdot e^{3t}\cdot 3$$

$$p'(t)=\frac{2e^{3t}}{\sqrt{4t+1}}+3e^{3t}\sqrt{4t+1}$$

$$p'(0)=\frac{2e^{0}}{\sqrt{4(0)+1}}+3e^{0}\sqrt{4(0)+1}=2+3=5$$

24. $x=\frac{3}{10}$

If $y=\ln(5x^2-3x+1)$, then $\frac{dy}{dx}=\frac{10x-3}{5x^2-3x+1}$.

$$\frac{10x-3}{5x^2-3x+1}=0 \;\Rightarrow\; 10x-3=0$$

25. The first derivative of $y=5^{(x^2)}$ is $\frac{dy}{dx}=5^{(x^2)}\ln(5)\cdot 2x$.

The second derivative is $\frac{d}{dx}\left(\frac{dy}{dx}\right)=2\ln(5)\frac{d}{dx}[x\cdot 5^{(x^2)}]$.

$$\frac{d^2y}{dx^2}=2\ln(5)[x\cdot 5^{(x^2)}\ln(5)\cdot 2x+5^{(x^2)}\cdot 1]=2\cdot 5^{(x^2)}\ln(5)[2x^2\ln(5)+1]$$

26. Use logarithmic differentiation.

$$\ln(y)=\ln\left(x^{\tan(x)}\right)=\tan(x)\cdot\ln(x)$$

$$\frac{1}{y} \cdot \frac{dy}{dx} = \tan(x) \cdot \frac{1}{x} + \ln(x) \cdot \sec^2(x)$$

$$\frac{dy}{dx} = x^{\tan(x)} \left[\frac{\tan(x)}{x} + \ln(x) \cdot \sec^2(x) \right]$$

27. Given $g(x) = \frac{\pi}{3}$, $\cos^{-1}(x^2) = \frac{\pi}{3} \Rightarrow x^2 = \cos\left(\frac{\pi}{3}\right)$.

$$x^2 = \frac{1}{2} \Rightarrow x = \frac{\sqrt{2}}{2}$$

$$g'(x) = \frac{-2x}{\sqrt{1-(x^2)^2}} = \frac{-2x}{\sqrt{1-x^4}}$$

$$g'\left(\frac{\sqrt{2}}{2}\right) = \frac{-2 \cdot \frac{\sqrt{2}}{2}}{\sqrt{1-\left(\frac{\sqrt{2}}{2}\right)^4}} = \frac{-\sqrt{2}}{\sqrt{1-\frac{1}{4}}}$$

$$\frac{-\sqrt{2}}{\sqrt{1-\frac{1}{4}}} = \frac{-\sqrt{2}}{\sqrt{\frac{3}{4}}} = \frac{-2\sqrt{2}}{\sqrt{3}}.$$

28. Use the product rule.

$$\frac{dy}{dx} = x^2 \cdot \frac{-1}{|x|\sqrt{x^2-1}} + 2x\csc^{-1}(x) = \frac{-x^2}{|x|\sqrt{x^2-1}} + 2x\csc^{-1}(x)$$

29. The derivative consists of the three heavier segments on the graph in the figure. Note that the derivative graph has positive values where $p(x)$ is increasing and negative values where $p(x)$ is decreasing.

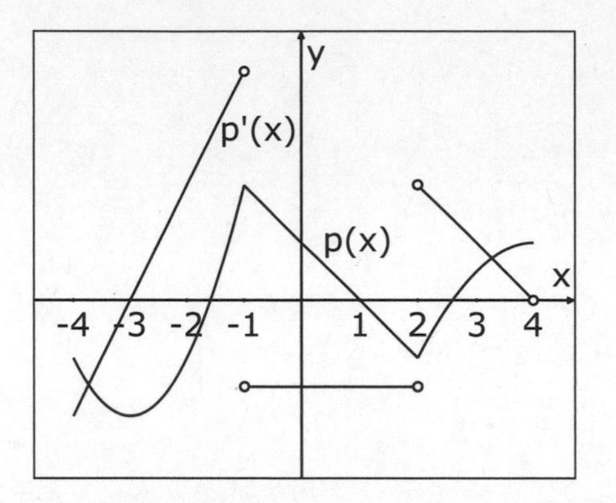

30. $f'(x) = 3x^2 + 10x - 8 = 0$ and $f''(x) = 6x + 10$

$(3x-2)(x+4) = 0 \ \Rightarrow \ x = \frac{2}{3}$ and $x = -4$ are critical points.

$f''\left(\frac{2}{3}\right) = 4 + 10 > 0$, which means that $f(x)$ has a local minimum at $x = \frac{2}{3}$.

$f''(-4) = -24 + 10 < 0$, which means that $f(x)$ has a local maximum at $x = -4$.

31. Find where the second derivative changes signs.

$$h'(x) = \frac{e^x \cdot 2x - (x^2+1) \cdot e^x}{e^{2x}} = \frac{(-x^2+2x-1)}{e^x}$$

$$h''(x) = \frac{e^x \cdot (-2x+2) - (-x^2+2x-1) \cdot e^x}{e^{2x}} = \frac{x^2-4x+3}{e^x}$$

$$\frac{x^2-4x+3}{e^x} = \frac{(x-1)(x-3)}{e^x} = 0 \ \Rightarrow \ x = 1 \ \text{ and } \ x = 3$$

Because $e^x > 0$, $h''(x)$ will change signs at $x = 1$ and $x = 3$, making both of these points inflection points.

32. Let r be the radius of the base, and let h be the height of the cylinder.

$SA = \pi \cdot r^2 + 2\pi rh = 16$

Isolate h to substitute into the volume equation. $h = \frac{8}{\pi \cdot r} - \frac{r}{2 \cdot \pi}$

$$V = \pi \cdot r^2 h = \pi \cdot r^2 \left(\frac{8}{\pi \cdot r} - \frac{r}{2 \cdot \pi} \right)$$

$$V = \pi r^2 h = \pi r^2 \left(\frac{8}{\pi r} - \frac{r}{2\pi} \right) = 8r - \frac{1}{2} r^3$$

$$\frac{dV}{dr} = 8 - \frac{3}{2} r^2 = 0$$

$$r^2 = \frac{16}{3} \quad \Rightarrow \quad r = \frac{4}{\sqrt{3}} \text{ inches}$$

33. If (x,y) is any point on the graph, and w is the straight-line distance from the origin to (x,y), then by the Pythagorean Theorem, $x^2 + y^2 = w^2$. At (2,4), $w = \sqrt{20}$.

Because the point is on $y = x^2$, substitute to get $y + y^2 = w^2$.

$$\frac{dy}{dt} + 2y \frac{dy}{dt} = 2w \frac{dw}{dt}$$

$$\frac{dy}{dt} + 2(4) \frac{dy}{dt} = 2\sqrt{20} \cdot 9$$

$$9 \frac{dy}{dt} = 18\sqrt{20} \quad \Rightarrow \quad \frac{dy}{dt} = 2\sqrt{20} = 4\sqrt{5} \text{ units per second}$$

34. $s'(x) = 6x^2$

$$s'(c) = \frac{s(2) - s(0)}{2 - 0}$$

$$6c^2 = \frac{15 - (-1)}{2} = 8$$

$$c^2 = \frac{4}{3} \quad \Rightarrow \quad c = \frac{2}{\sqrt{3}}$$

The negative solution is not included because it is not in the interval $[0,2]$.

35. $\Delta x = \dfrac{\pi - 0}{6} = \dfrac{\pi}{6}$

$$\text{LRAM}_6 = \frac{\pi}{6}\left[g(0) + g\left(\frac{\pi}{6}\right) + g\left(\frac{\pi}{3}\right) + g\left(\frac{\pi}{2}\right) + g\left(\frac{2\pi}{3}\right) + g\left(\frac{5\pi}{6}\right)\right]$$

$$\text{LRAM}_6 = \frac{\pi}{6}\left[4 + \frac{7}{2} + \frac{5}{2} + 2 + \frac{5}{2} + \frac{7}{2}\right] = \frac{\pi}{6}(18) = 3\pi$$

36. $h = \Delta x = \dfrac{6-1}{5} = 1$

Let $p(x) = \sqrt{x+1}$.

$$\text{TRAP}_5 = \frac{1}{2}[p(1) + 2p(2) + 2p(3) + 2p(4) + 2p(5) + p(6)]$$

$$\text{TRAP}_5 = \frac{1}{2}\ \sqrt{2} + 2\sqrt{3} + 2\sqrt{4} + 2\sqrt{5} + 2\sqrt{6} + \sqrt{7}\ \approx 10.448$$

37. $\displaystyle\int_1^6 \sqrt{x+1}\,dx = \int_1^6 (x+1)^{(1/2)}\,dx$

$$\int_1^6 (x+1)^{(1/2)}\,dx = \frac{2}{3}(x+1)^{(3/2)}\Big|_1^6 = \frac{2}{3}\left[7^{(3/2)} - 2^{(3/2)}\right]$$

The error is approximately $\dfrac{2}{3}\left[7^{(3/2)} - 2^{(3/2)}\right] - 10.448 = 0.01363$.

38. $h(x)$ lies below the x-axis for $0 \le x < 2$ and above the x-axis for $2 < x \le 3$. Two integrals are needed.

$$-\int_0^2 (2^x - 4)\,dx + \int_2^3 (2^x - 4)\,dx = -\left(\frac{2^x}{\ln(2)} - 4x\right)\Bigg|_0^2 + \left(\frac{2^x}{\ln(2)} - 4x\right)\Bigg|_2^3$$

$$-\left(\frac{2^x}{\ln(2)} - 4x\right)\Bigg|_0^2 + \left(\frac{2^x}{\ln(2)} - 4x\right)\Bigg|_2^3 = \left(8 - \frac{3}{\ln(2)}\right) + \left(\frac{4}{\ln(2)} - 4\right) = 4 + \frac{1}{\ln(2)}$$

39. By the Fundamental Theorem, $a'(x) = f(x)$ and $a''(x) = f'(x)$.

Inflection points on $a(x)$ happen where $a''(x) = f'(x)$ changes sign. On the graph of $f(x)$, this is where the slope of $f(x)$ changes sign.

Inflection points are at $x = -3$, $x = 0$ and $x = 2$.

40. Using the Fundamental Theorem, $k'(x) = \frac{\sqrt[3]{x^2 - 1}}{x^2 + 2} \cdot 2x$.

$$k'(3) = \frac{\sqrt[3]{3^2 - 1}}{3^2 + 2} \cdot 2(3) = \frac{12}{11}$$

41. Let $u = x^2 + 4x + 5$ and $du = (2x + 4)\,dx$ or $\frac{1}{2}du = (x + 2)\,dx$.

$$\int \frac{3(x+2)}{x^2 + 4x + 5}\,dx = \int \frac{3 \cdot \frac{1}{2}du}{u}$$

$$\frac{3}{2}\int \frac{du}{u} = \frac{3}{2}\ln|u| + C = \frac{3}{2}\ln|x^2 + 4x + 5| + C$$

42. Choice C, $\frac{dy}{dx} = x^{\left(-\frac{1}{3}\right)}$, is the correct choice. Recognizing that $x^{(-1/3)} = \frac{1}{\sqrt[3]{x}}$ makes it easier to analyze the slopefield. For $x < 0$ all slopes are negative, and for $x > 0$ all slopes are positive. As the absolute value of x increases, the absolute values of the slopes decrease. It cannot be choice A, because the slopes at $x = 1$ should be undefined. It cannot be choice B or choice D, because all slopes for those differential equations are positive.

43. For $\frac{dy}{dx} = 4x\sqrt{y}$, a wise separation of variables is $\frac{dy}{2\sqrt{y}} = 2x\,dx$.

$$\int \frac{dy}{2\sqrt{y}} = \int 2x\,dx$$

$$\sqrt{y} = x^2 + C$$

Using $y(1) = 9$, $\sqrt{9} = 1^2 + C \;\Rightarrow\; C = 2$

$$\sqrt{y} = x^2 + 2 \quad \text{or} \quad y = (x^2 + 2)^2$$

44. $\frac{1}{7-1}\int_1^7 \frac{4}{2x+1}dx$

Let $u = 2x+1$ and $du = 2dx$. Then when $x=1$, $u=3$, and when $x=7$, $u=15$.

$$\frac{1}{7-1}\int_1^7 \frac{4}{2x+1}dx = \frac{1}{6}\int_3^{15}\frac{2du}{u}$$

$$\frac{2}{6}\int_3^{15}\frac{du}{u} = \frac{1}{3}\ln(u)\Big|_3^{15} = \frac{1}{3}[\ln(15)-\ln(3)]$$

45. Velocity is negative for $0 \le t < 1$ and positive for $1 < t \le 8$, so two integrals are necessary.

$$-\int_0^1 [t^{(2/3)}-1]dt + \int_1^8 [t^{(2/3)}-1]dt = -\left[\frac{3}{5}t^{(5/3)}-t\right]\Bigg|_0^1 + \left[\frac{3}{5}t^{(5/3)}-t\right]\Bigg|_1^8$$

$$-\left[\frac{3}{5}t^{(5/3)}-t\right]\Bigg|_0^1 + \left[\frac{3}{5}t^{(5/3)}-t\right]\Bigg|_1^8 = -\left[\frac{3}{5}-1\right]+\left[\left(\frac{3}{5}8^{(5/3)}-8\right)\right]-\left[\left(\frac{3}{5}1^{(5/3)}-1\right)\right]$$

$$-\left[\frac{3}{5}-1\right]+\left[\left(\frac{3}{5}8^{(5/3)}-8\right)\right]-\left[\left(\frac{3}{5}1^{(5/3)}-1\right)\right] = \frac{2}{5}+\frac{96}{5}-8-\left(\frac{-2}{5}\right) = 12 \text{ feet.}$$

46. The parabola is above the cubic, and the curves intersect at (2, 3).

$$\text{Area} = \int_0^2 \left(5-\frac{1}{2}x^2\right)-\left(1+\frac{1}{4}x^3\right)dx = \int_0^2\left(4-\frac{1}{4}x^3-\frac{1}{2}x^2\right)dx$$

$$\int_0^2\left(4-\frac{1}{4}x^3-\frac{1}{2}x^2\right)dx = 4x-\frac{1}{16}x^4-\frac{1}{6}x^3\Big|_0^2$$

$$4x-\frac{1}{16}x^4-\frac{1}{6}x^3\Big|_0^2 = 8-\frac{1}{16}\cdot 16-\frac{1}{6}\cdot 8 = \frac{17}{3}$$

47. $V = \pi\int_0^\pi [m(x)]^2\,dx = \pi\int_0^\pi\left[\sqrt{\sec\left(\frac{x}{3}\right)\tan\left(\frac{x}{3}\right)}\right]^2 dx$

$$\pi\int_0^\pi \sec\left(\frac{x}{3}\right)\tan\left(\frac{x}{3}\right)dx = \pi\cdot 3\sec\left(\frac{x}{3}\right)\Bigg|_0^\pi$$

$$\pi \cdot 3\sec\left(\frac{x}{3}\right)\Big|_0^{\pi} = 3\pi\left[\sec\left(\frac{\pi}{3}\right) - \sec(0)\right] = 3\pi(2-1) = 3\pi$$

48. There is space between the region and the axis of rotation, so the washer method is needed.

$$\pi\int_1^2 (3x-2)^2 - (x^x)^2\, dx = \pi\int_1^2 9x^2 - 12x + 4 - x^{2x}\, dx$$

49. The area of a square can be found by $A = \frac{1}{2}d^2$, where d is the length of the diagonal. For each cross section, the diagonal is determined by the given function.

$$V = \int_0^2 [A(x)]dx = \frac{1}{2}\int_0^2 \left[e^{(-x^2)}\right]\cdot\sqrt{x}^{\,2}\, dx = \frac{1}{2}\int_0^2 x\cdot e^{(-2x^2)}\, dx$$

Let $u = -2x^2$ and $du = -4x\,dx$. If $x = 0$, then $u = 0$. If $x = 2$, then $u = -8$.

$$\frac{1}{2}\int_0^2 x\cdot e^{(-2x^2)}\, dx = \frac{1}{2}\int_0^{-8} e^u\cdot\left(-\frac{1}{4}\right)du = -\frac{1}{8}\int_0^{-8} e^u\, du$$

$$-\frac{1}{8}\int_0^{-8} e^u\, du = -\frac{1}{8}e^u\Big|_0^{-8} = -\frac{1}{8}[e^{-8} - e^0] = \frac{1}{8} - \frac{1}{8}e^{-8}$$

50. Use the formula for the length of an arc: $L = \int_a^b \sqrt{1 + (f'(x))^2}\, dx$.

$$c'(x) = \frac{1}{2}(x^2+1)^{(-1/2)}\cdot 2x = \frac{x}{\sqrt{1+x^2}}$$

$$[c'(x)]^2 = \frac{x^2}{1+x^2} \text{ and } 1 + [c'(x)]^2 = 1 + \frac{x^2}{1+x^2} = \frac{1+2x^2}{1+x^2}$$

$$L = \int_{-1}^3 \sqrt{\frac{1+2x^2}{1+x^2}}\, dx \approx 4.858$$

Index

Note: Page numbers in *italics* indicate answers to problems. For problems on specific topics, reference the chapter where the topic falls in the *Skill Check problems* main entry.

C

D

H

I

J

L

M

N

O

P